Mineral Processing Design

NATO ASI Series

Advanced Science Institutes Series

A Series presenting the results of activities sponsored by the NATO Science Committee, which aims at the dissemination of advanced scientific and technological knowledge, with a view to strengthening links between scientific communities.

The Series is published by an international board of publishers in conjunction with the NATO Scientific Affairs Division

A	Life Sciences	Plenum Publishing Corporation
B	Physics	London and New York
C	Mathematical and Physical Sciences	D. Reidel Publishing Company Dordrecht, Boston, Lancaster and Tokyo
D	Behavioural and Social Sciences	Martinus Nijhoff Publishers Boston, Dordrecht and Lancaster
E	Applied Sciences	
F	Computer and Systems Sciences	Springer-Verlag Berlin, Heidelberg, New York
G	Ecological Sciences	London, Paris, Tokyo
H	Cell Biology	

Series E: Applied Sciences – No. 122

ML

Mineral Processing Design

Editors:

B. Yarar

Colorado School of Mines
Department of Metallurgical Engineering
Golden, CO 80401, USA

Z.M. Dogan

Middle East Technical University
Department of Mining Engineering
Ankara, Turkey

1987 **Martinus Nijhoff Publishers**
Dordrecht / Boston / Lancaster
Published in cooperation with NATO Scientific Affairs Division

Proceedings of the NATO Advanced Study Institute on Mineral Processing Design,
Bursa, Turkey, August 20-31, 1984

NATO Advanced Study Institute on Mineral Processing
    Design (1984 : Bursa, Turkey)
    Mineral processing design.

    (NATO advanced science institutes. Series E,
Applied sciences ; no. 122)
    "Proceedings of the NATO Advanced Study Institute
on Mineral Processing Design, Bursa, Turkey, August
20-31, 1984"--T.p. verso.
    Includes bibliographies and index.
    1. Ore-dressing--Congresses. 2. Mines and mineral
resources--Congresses. I. Yarar, Baki. II. Dogan,
Z.M. III. Title. IV. Series.
TN500.N26 1984   622'.7   86-33126

ISBN 90-247-3472-X (this volume)
ISBN 90-247-2689-1 (series)

Distributors for the United States and Canada: Kluwer Academic Publishers,
P.O. Box 358, Accord-Station, Hingham, MA 02018-0358, USA

Distributors for the UK and Ireland: Kluwer Academic Publishers, MTP Press Ltd,
Falcon House, Queen Square, Lancaster LA1 1RN, UK

Distributors for all other countries: Kluwer Academic Publishers Group, Distribution
Center, P.O. Box 322, 3300 AH Dordrecht, The Netherlands

Printed in The Netherlands

TABLE OF CONTENTS

Foreword

This volume is based on the proceedings of the "NATO-
Advanced Study Institute on Mineral Processing Design" held in
Bursa-Turkey on August 24-31, 1984.

The institute was organized by Professor B. Yarar of the
Colorado School of Mines, Golden, Colorado, 80401, USA, Professor
G. Ozbayoghu and Professor Z. M. Dogan of METU-Ankara, Turkey,
who was the director.

The purpose of the institute was to provide an international
forum on the subject and update the information available.

Participants were from Turkey, England, Greece, Spain,
Portugal, Belgium, Canada, and the USA.

Besides authors contributing to this volume, presentations
were also made by Drs. Yarar, Raghavan, Schurger, and Mr. Kelland.

Many assistants and colleagues helped. They are gratefully
acknowledged. Acknowledgment is also owed to Drs. Ek, de Kuyper,
and Tolun.

Dr. Gûlhan Ozbayoglu, and Mr. S. Ozbayoglu were particularly
helpful in the overall organization and hosting of many
international guests. We owe them special thanks.

NATO, Scientific Affairs Division, is gratefully
acknowledged for the grant which made this activity possible.

Z. M. Dogan
B. Yarar

2

APPLIED MINERALOGY IN ORE DRESSING

William Petruk

CANMET, 555 Booth Street, Ottawa, Ontario, K1A 0G1

ABSTRACT

    Mineralogy applied to ore dressing is a reliable guide for
designing and operating an efficient concentrator. A procedure
for conducting mineralogical studies in conjunction with ore
dressing was, therefore, developed. The procedure includes
characterizing the ore and analysing the mill products.
Characterizing the ore involves identifying the minerals,
determining the mineral quantities, and measuring the size
distributions of the minerals in the uncrushed ore. The size
distribution data are incorporated into liberation models to
predict mineral liberations in screened fractions. The mill
products are analysed by studying polished sections to define the
behaviour of the minerals in the concentrator. This involves
measuring the liberation and size distribution for each mineral
and characterizing the middling particles. The quantities of free
and unliberated mineral grains in the concentrate and tailings
from each circuit define the behaviour of the minerals and show
whether poor metal recoveries are due to poor concentrator
performance or to the nature of the ore.

1. INTRODUCTION

A fundamental requirement in designing a processing plant for high metal recoveries is a thorough understanding of the mineralogical characteristics of the ore. An approach to conducting mineralogical studies of ores in conjunction with ore dressing was developed at CANMET, and is used as a guide for performing mineralogical studies [1,2]. The approach considers the objectives of both ore dressing and applied mineralogy. The objective in an ore dressing operation is to concentrate particles of specific minerals and to obtain satisfactory concentrate grades and metal recoveries. To concentrate the minerals, the mineral particles must either be free or contain only small amounts of other minerals. Minerals are liberated by reducing the sizes of the ore pieces in stages: first to about minus 1 cm with a crusher, then to minus 10 mesh (1.65 mm) with a rod mill, and finally to free particles by using one or more ball mills in closed circuit with cyclones.

The objective of applied mineralogy related to ore dressing is to provide guidance for grinding the ore and for concentrating the mineral particles of interest. This is done by first studying the uncrushed ore and then studying the mill products from selected sites in a mill.

2. STUDIES ON UNCRUSHED ORES

Polished sections of uncrushed ore samples are studied prior laboratory bench tests to establish preliminary conditions for testing and to obtain a preliminary assessment of possible metal recoveries. This study involves

(1) identifying the major, minor and trace minerals

(2) allocating each element to each mineral

(3) determining the quantities of each mineral, and

(4) determining the grain size distributions of the minerals of economic value.

The samples for study should reflect mineralogical variations of the orebody keeping in mind that the concentrator will likely be treating blended ore. During this study chemical and spectrochemical analyses should be available as a reference. The study could be general or extensive. An extensive study is preferential, although a general study can provide useful information. For example a cursory study of the New Brunswick

Zn-Pb-Cu-Ag ores has shown that the minerals of economic value are sphalerite, galena, chalcopyrite and tetrahedrite and that they occur as small inclusions in massive pyrite. Minor amounts of arsenopyrite, magnetite and pyrrhotite are present, and the sphalerite contains small inclusions of cassiterite. This cursory study has served to identify the main host minerals for Zn, Pb, Cu, Sn and probable host minerals for the Ag and Bi. An extensive study with an electron microprobe, a scanning electron microscope(SEM) and an image analyser had to be carried out to obtain the necessary information for allocating the silver to its host minerals and for calculating a mineral balance for silver. The extensive study established that the main silver-bearing minerals are tetrahedrite and galena, and small amounts of the silver occur in pyargyrite and acanthite. The tetrahedrite contains about 16 wt % Ag and the galena about 0.1 wt % Ag [3]. The cursory study has also shown that the ore is fine-grained and will have to be ground finely to liberate the minerals. An extensive study with an image analyser determined the grain size distribution in the ore and enabled predicting the minimum and optimum grinds for liberating the minerals.

A discussion on methods of mineral identification is given in a later section of this paper.

Information on mineral quantities can be determined by point counting, by X-ray diffractometer analysis combined with chemical analysis, and by image analysis. The basic principle in the point counting technique is that the relative volume % of different classes of individuals is equal to the number of times that each individual is encountered under a cross hair when the section is moved at equal intervals across the field. In image analysis either surface areas or chord lengths are measured. The basic principle here is that the relative volume % of different classes of individuals is equal to the relative surface areas covered by all individuals, or to the relative chord lengths that intersect the individuals provided that the section is so traversed that each grain is intersected only once by the chord.

Grain size distributions of the ore minerals in an uncrushed ore are essential for predicting mineral liberations at particular grinds. The size distributions can be determined manually by point counting, or automatically by image analysis. Many mathematical models have been proposed for predicting mineral liberations on the basis of grain size distributions in the uncrushed ore. Some are based on manual point counting [4,5], some on measuring linear intercepts [6], and others on measuring surface areas [7,8]. Each model predicts the mineral liberation

that would be obtained in each screened fraction of the ground
ore. Image analysis studies using the model of Petruk [8] showed
that the minimum grind for the sphalerite in the ore from
Brunswick Mining and Smelting is 80% minus 200 mesh and optimum
grind is 500 mesh. The galena and chalcopyrite are finer grained,
hence the optimum grind is about 80% minus 15 µm. Grinding tests
have confirmed this evaluation. The author has found empirically
that the minimum grind is the grind at which about 50% of the
minerals are as free grains. This generally occurs when the size
distribution of the ground particles is equal to the size
distribution of the mineral in the uncrushed ore.

3. STUDIES OF CRUSHED MILL PRODUCTS

After a preliminary assessment has been made by studying
fragments of the ore it is necessary to study mill products in
order to:

(1) evaluate the actual response of the ore to ore dressing;

(2) assess the plant performance; and

(3) establish whether plant modifications are needed to improve
    metal recoveries.

The main objective in the plant is to obtain high recoveries
and concentrate grades that are acceptable for subsequent
processing. This can only be achieved by plants that are designed
for recovering free mineral grains. Therefore plant performance
can be assessed by determining the recovery of free mineral
grains. When the plant is performing properly some unliberated
mineral grains will also be recovered, the amount and variety of
particle will be dependent upon the ore characteristics. One
must, therefore, determine the plant performance and response of
the ore to processing at the same time by:

(1) performing metal and mineral balances for the concentrates
    and tailings;

(2) determining mineral liberations in the various mill products;

(3) determining recoveries of the free mineral grains;

(4) determining recoveries and behaviours of the unliberated
    mineral grains; and

(5) performing mineralogical investigations related to special
    recovery problems.

6

3.1 Metal and Mineral Balances

The term metal balance as used in this paper will refer to the distribution of metals at different points in a concentrator at a particular time, and the term mineral balance will refer to the distribution of minerals. It is generally believed by mineralogists that a metal balance can be obtained by multiplying the assay of each product X the weight of the product and calculating the metal distribution. In practice, due to sampling, assaying, weighing and other errors, the assay values X the weights for all products from a process are never equal to the assay of the feed X the weight of the feed to the process. Therefore, a material balance adjustment [9] analysis to adjust the assay values and the weights for best fit must be performed.

To assess the response of an ore to a concentrating process it is necessary to obtain a mineral balance because the concentrator is designed to concentrate minerals. This is done by:

(1) assaying each product and performing a metal balance;

(2) identifying the metal-bearing minerals. If only one mineral is present for each metal, the mineral balance is equal to the metal balance which can now be evaluated;

(3) if the metal is present in more than one mineral it is necessary to

(a) determine the average quantity of metal in each mineral

(b) determine the quantity of each mineral in each product

(c) perform a material balance which will adjust the mineral distribution to a mineral balance

3.2 Mineral Liberation

Mineral liberation can be assessed with polished sections by measuring the surface areas of free and unliberated mineral grains. The observed proportions of free mineral grains are always larger than the actual proportions of free mineral grains because many unliberated grains may be so oriented that they appear free in polished sections [4,10]. If one measures the proportion of free grains in a series of sized fractions of a mill product the values generally will not plot as a smooth curve because anomalously high readings will be obtained (Fig. 1). It is likely that the actual proportions of free mineral grains in each sized fraction will be close to a curve drawn through the lowest points as shown in Fig. 1. It may be possible to make appropriate stereological corrections to account for the high readings but procedures for making the corrections are not known. Mineral liberation data of this type can be used by ore dressing engineers to determine the approximate minimum and optimum grinds and mineral liberation in the sample. The minimum grind is the point where the curve begins to steepen and optimum grind where it begins to flatten. For the example in Fig. 1 the minimum grind is about 100% minus 125 μm or approximately 80% minus 75 μm, and the optimum grind is about 100% minus 20 μm or approximately 80% minus 15 μm. The expected mineral liberation in the sample can be calculated using the weight of each screen fraction, the per cent free in each screened fraction, and the assay for the element.

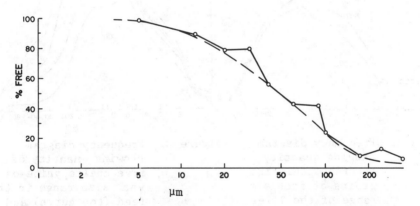

Figure 1. Curve showing percent of mineral as free grains in a series of polished sections of screened grains from one sample. Solid line represents observed values; dashed line represents interpreted quantities of free grains.

3.3 Free Grains

In order to obtain high recoveries the minerals must be present as free or nearly free grains [11] and the mill must operate efficiently. Low metal recoveries are, therefore, due to either poor mill performance or poor mineral liberation. An indication of poor mill performance may be obtained by examining

the mill products with an ore microscope [12]. A definite
assessment of mill performance, however, involves determining the
recovery of free mineral grains in the concentrate and losses in
the tailings and other concentrates (Fig. 2 and 3). Such an
assessment will show whether the overall recovery of free grains
is satisfactory, as in Fig. 2 where the recovery of free
sphalerite was 86% [3]. An example of an unsatisfactory recovery
of free grains is shown in Fig. 3 where only 49% of the free
chalcopyrite was recovered in the copper concentrate. The main
loss was in the lead concentrate and tailings as large free grains
[3]. The ore dressing engineers used the information obtained for
the examples in Fig. 2 and 3 very profitably. In the case of
sphalerite, they discontinued further attempts to fine-tune the
concentrator, whereas, in the case of chalcopyrite, they conducted
new flotation tests and improved the copper recovery.

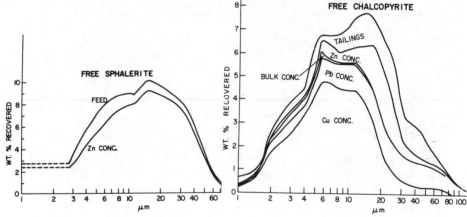

Figure 2. Frequency diagram
showing quantity
of free sphalerite
grains of each size
range of the Tyler
Series in the feed
(top curve) and in
zinc concentrate
(bottom curve). The
space between the
curves represents the
quantity and size of
free sphalerite lost
to the tailings and
other concentrates.

Figure 3. Frequency diagram
showing quantity of
free chalcopyrite of
each size range in the
feed (top curve) and
copper concentrate
(bottom curve). The
space between succes-
sive curves represents
the quantity and size
of free chalcopyrite
lost to the tailings
and other concen-
trates. The total %
age free chalcopyrite
for all size ranges in
the feed is 100%.

3.4   Unliberated Grains

Middling particles are composed of grains of the desired
mineral intergrown with grains of one or more other minerals.
Such particles pose a problem in ore dressing because some have
characteristics that cause them to be drawn into the concentrate,
some are rejected to the tailings, and some have intermediate
characteristics that may cause them to go either way. If too many
middling particles are in the concentrate the grade will be low,
if too many are in the tailings the recovery will be low. In
studying mill products it is necessary to define a classification
for the middling particles and to establish the classes of
particles that tend to be directed to either the concentrates or
tailings. Furthermore, it is necessary to determine the quantity
of intermediate grains. If there is a significant amount of
intermediate middling particles they must be collected as a
scavenger concentrate and reground. The significance of the
middling particles in ore dressing was recognized by Amstutz [13]
who classified the textural intergrowth of the particles and
correlated the classification with mineral recoveries. His
classification is useful for evaluating mill products when
studying them by ore microscopy and is widely used [14].

In evaluating middling particles by automatic image analysis
[15] a different approach has to be used because the image
analyser cannot recognize textures as readily as the human eye.
On the other hand precise measurements of the surface areas
covered by grains in polished section can be made, and thereby the
percentage of the desired mineral in each middling particle, and
the grain sizes of the desired mineral and of the middling
particle can be determined [16]. The middling particles can be
arranged in classes containing 0-10, 10-20, 20-30, etc. vol. %
desired mineral. The quantity of desired mineral in the middling
particle in each class would then be measured (Fig. 4). The
classes of middling particles that tend to enter the concentrate
or tailings can be established by analyzing the middling particles
in each or interpreted by using mineral flotabilities [4]. When
this is established the middling particles in the ore are analysed
to determine the quantity and size of middling particle in each
class. From these data the cumulative liberation yield, proposed
by Miller et al. [17], can be calculated and used to define the
optimum recovery of the desired mineral in the concentrate and the
expected loss to the tailings. The cumulative liberation yield is
the proportion of a mineral that is present as free grains plus
the proportion in middling particles in each succeeding class.
For example, in one study (Fig. 4) the average middling particles
in the zinc concentrate contained 68 vol. % sphalerite. The
distribution of sphalerite in the ore was 66% as free grains and

24% in middling particles containing more than 68 vol. % sphalerite. The <u>cumulative liberation yield</u> for sphalerite in a zinc concentrate containing particles with more than 68 vol. % sphalerite was 66+24=90%. Actual recovery was 83% Zn in the zinc concentrate plus 9% Zn in the Pb concentrate. This shows that <u>cumulative liberation yield</u> is a practicle assessment of the actual plant performance.

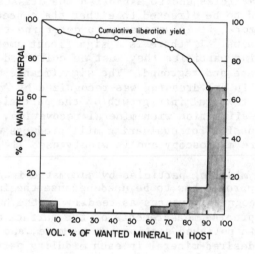

Figure 4. Typical example of a histogram that shows 10 classes of middling particles based on vol. % of wanted mineral in each particle. The height of each bar gives the % of wanted mineral within each class of middling particle. The sum of wanted mineral in all classes is 100%. Cumulative liberation yield = quantity of wanted mineral in 90-100% class plus quantity of wanted mineral in each succeeding class. The 90-100% class is considered as free grains.

     Size analyses of the desired and middling grains are useful refinements. For example, it has been found that middling grains containing sphalerite inclusions smaller than 10 μm and larger than 40 μm tend to be dropped into the tailings, whereas middling grains containing sphalerite inclusions 10-40 μm in diameter tend to be pulled into the concentrate. The presence in specific mill products, of unusual quantities of middling particles with too

large or too small inclusions could indicate that the plant
operation needs to be modified.

4. APPLIED MINERALOGY IN CIRCUITS TREATING VOLCANOGENIC BASE
    METAL ORES

To effectively monitor a flotation concentrator it is
necessary to know the behaviour of the mineral in various parts of
the concentrator, to understand the reason for the mineral
behaviour, and to know what can be done to reduce losses.

4.1 Copper-Lead Circuit

The first circuit in nearly all concentrators that treat
volcanogenic base metal ores by flotation is the primary Cu-Pb
circuit. The circuit recovers chalcopyrite, galena, many silver
bearing minerals, stannite, gold inclusions in chalcopyrite and
galena and other minerals. The circuit tailings contain the
sphalerite, pyrite, non-metallic minerals, slow-floating
tetrahedrite, some of the silver sulfosalts [18] and minute
inclusions of gold in pyrite.

Specific conditions that cause problems in the primary Cu-Pb
circuit are:

(1) SO_2, which is often used as a conditioning agent in the
     grinding circuit, depresses fine-grained acanthite at a high
     pH ($\approx$ 9). The mineral therefore, tends to be lost to
     tailings.

(2) In some plants the concentrate from the primary Cu-Pb circuit
     tends to selectively recover mineral grains of a size range
     that is unique for each circuit. If the circuit is
     recovering small grains (2-15 μm) then oversize (25 μm) free
     grains will be lost to the tailings [3] [4].

(3) The presence of about 0.01 wt % or more of covellite or
     chalcocite affects the ore in such a manner that chalcopyrite
     and galena cannot be selectively floated from sphalerite
     [19]. This problem can be partly overcome by adding NaCN to
     the floation circuit. The sphalerite, however, is depressed
     only when an excess of NaCN has been added and free CN is
     present. There must, therefore, be enough NaCN to dissolve
     all of the covellite and chalcocite. Hence, the technique is
     not economical when the ore contains a significant amount of
     these minerals (more than 0.1 wt % contained Cu).

4.2   Copper-Lead Separation Circuit

The concentrate from the Cu-Pb primary circuit is passed through a copper-lead separation circuit which recovers the chalcopyrite, fast-floating tetrahedrite, pyrargyrite, and stannite in a copper concentrate. Inclusions of native gold and electrum are also recovered. The concentrate is a saleable product and payment is received for the copper, silver and gold [20].

4.3   Lead Circuit

The tailing from the copper-lead separation circuit is directed to a lead circuit which recovers the galena, chalcopyrite, silver and bismuth sulphosalts, remaining fast-floating tetrahedrite, gold inclusions in galena and other minerals. The lead concentrate from this circuit is generally impure but it is a saleable product and payment is received for the contained lead, silver and gold. The silver in this concentrate occurs in solid solution in galena, in tetrahedrite and in the silver sulfosalts. Some bismuth, copper, arsenic and antimony, with trace amounts of selenium and tellurium [21] may also be present in the concentrate. The arsenic is in arsenopyrite and in solid solution in any pyrite that was drawn into the concentrate as part of middling galena-pyrite grains. The antimony is in tetrahedrite, and the selenium and tellurium occur, probably in trace quantities, in tetrahedrite and chalcopyrite.

4.4   Zinc Circuit

The tailing from the primary Cu-Pb circuit and from the lead circuit is directed to a zinc circuit where a zinc concentrate is produced. The zinc concentrate generally contains 50 to 400 g/tonne Ag, some In, Cd, Mn, and Au. About 10 to 20% of the silver from many volcanogenic ores is in the zinc concentrate but generally no payment is received for this silver. The silver in such zinc concentrates is contained in the galena, in tetrahedrite, and frequently ($\approx$40% of the Ag in the zinc concentrate) in other minerals such as acanthite and native silver [22]. The acanthite and native silver are so fine grained ($\approx$0.1 μm) that they cannot be removed from the concentrate by normal flotation procedures. The silver contained in these minerals can, however, be recovered by cyanidation of the zinc concentrate. This has not yet been done in any plant because the concept has not been tested with a pilot plant. The In, Cd and Mn are contained in the sphalerite but generally no payment is received for these metals. Small amounts of Hg are generally present in the sphalerite. Gold that occurs as inclusion in sphalerite would be recovered in the zinc concentrate. Payment may be received for the contained gold.

The tailing from the zinc circuit is usually the plant tailing and it may contain small amounts of gold. In some instances the gold is present as minute inclusions in pyrite. In other instances, particularly where the gold content of the ore is less than 1 g/tonne, as for the ore from the Heath Steele deposit in New Brunswick, the gold-bearing mineral(s) have not been found nor identified by current laboratory techniques [18]. Assays of mineral concentrates prepared from the tailings indicate however, that most of the gold in the tailings is in pyrite.

4.5 Tin Circuit

Most of the tin in volcanogenic base metal ores occurs in cassiterite although some is in stannite. The stannite is generally associated with chalcopyrite and tends to be recovered in the copper concentrate. No payment is received for this tin.

The cassiterite generally occurs as small inclusions in sphalerite but, because it is much harder than the sphalerite, the small grains tend to be liberated during grinding. These grains are rejected from the zinc circuit and are directed to the tailings. Some plants have installed tin circuits using gravity techniques. Such techniques are inefficient for the very fine-grained minerals (1–25 μm), therefore, the tin recovery is poor. Since the cassiterite is so fine grained it is recommended that flotation be used to recover the mineral from volcanogenic base metal ores.

5. EXAMPLES OF BENEFICIATION PROBLEMS EXPLAINED BY MINERALOGICAL STUDIES

Examples of simple to complex ore beneficiation problems that have been encountered either in the plant or at the bench are given in this section. Mineralogical studies explained the reasons for the problems and in some cases resolved them.

5.1 Case I – Lead Concentrates at Fletcher Mill Southeast Missouri

The zinc contents of the lead concentrates are generally low, but at times the zinc content increases to 3.3 wt % Zn. A microscopical study of the lead concentrate showed that the sphalerite was present as free particles. The ore dressing engineer realized that this was due to a sloughing over of the lead cleaner concentrate. Closer control of the cleaner cell eliminated the problem [23].

5.2 Case II - Copper Concentrates from the Lead Belt Missouri

The lead content of the concentrates varies from 5 to 20 wt % and is too high for U.S. Cu smelters. Ore microscopy particle counting of the screen fractions of the Cu ore showed that the proporation of free to locked galena increases with decreasing particle size, and chemical analyses show that the largest amount of Pb is contained in the smallest size fraction. Further microscopical examinations showed that some fine galena is trapped in clusters of chalcopyrite particles. Increased agitation during lead copper separation has caused better separation of the free galena and chalcopyrite particles, and made it possible to market the Cu concentrate domestically [23].

5.3 Case III - Low Grade Copper Concentrate

The copper concentrate grade from one mill had decreased significantly. Polished section studies of the concentrate showed numerous grains of pyrite rimmed by digenite and covellite. The rimming caused the pyrite to behave in a similar manner to the digenite and covellite and was not being depressed. This diluted the concentrate grade [24].

5.4 Case IV - Poor Recovery of Cassiterite

A study was conducted to characterize the mineralization and to develop a process for recovering tin from the Bolivian tin deposits [24]. Cassiterite was the principle tin mineral and the host rock was altered quartz sandstone with a fine-grained interstitial cement composed mainly of quartz, mica and clay minerals. Much of the cassiterite was associated with the interstitial cement as clusters of intergrown crystals, but some of the cassiterite was coarse-grained. This required developing a flow sheet that uses gravity concentration of the coarse-grained and flotation of the fine cassiterite.

5.5 Case V - Alteration of Ni-Sulphide Ore

(a) Pentlandite - Violarite. In the Inco Nickel sulphide deposits altered sulphides are encountered in pockets and they have deleterious effects on sulphide flotation [25]. One type of alteration is the alteration of pentlandite to violarite plus magnetite. Associated with it is alteration of pyrrhotite to

pyrite, nickeliferous smythite, and magnetite. The secondary
pyrite has a high nickel content. Processing problems are related
to the textures produced by the alteration and no real solution is
apparent. There are two problems associated with this assemblage;
(a) on grinding, the violarite grains commonly retain the
associated magnetite veinlets as rims, and when too much magnetite
is exposed these grains are lost to the tailing. (b) The presence
of nickeliferous pyrite and smythite adversely affect the
production of Ni concentrates. Minimum tailing assay produced
from this ore was 0.33 wt % Ni whereas normal tailing is 0.16 wt %
Ni.

(b) Occurrence of Tochilinite. Tochilinite
$(6(Fe,Ni)S.5(Mg,Fe^{2+})(OH)_2)$ is a member of the valleriite mineral
series. It was found in a propsect from the Thompson area in
Manitoba where it occurs as inclusions in pentlandite and
pyrrhotite [25]. Poor results of mill tests were obtained from
the tochilinite ore. Chemical analyses of sized fractions of the
tailings showed that initially much of the nickel loss was in
50-100 μm size particles. This was due to tochilinite inclusions
in pentlandite. Regrinding and subsequent flotation reduced the
Ni content of the tailing by half. However, the level was still
high (0.44%, normal 0.16%). Microprobe and X-ray diffraction
studies show that pentlandite, pyrrhotite and tochilinite all
contribute to the tailing loss. As with the violarite problem no
solution is readily apparent: if tochilinite is rejected to the
tailing it contributes to a tailing loss, but if it is recovered
it reduces the concentrate grade.

5.6 Case VI - Submicroscopic Gold in Clay

An ore sample assaying 0.22 oz Au/ton contained quartz (49%),
illite (15%), kaolin (7%), calcite (8%), dolomite (17%), barite
(2%) and hematite (2%) [26]. Microscopical studies of the ore in
polished section revealed a single minute gold particle in the
clays. To establish the occurrence of gold, mineral concentrates
were prepared by (a) sedimenation and decantation, (b) heavy
liquids and (c) a Frantz Isodynamic separation. Slimes were
siphoned from 1000 ml cylinders after allowing to settle in
calgondeionized water solution for intervals of 30, 60 and 120
minutes. Most but not all of the clays were removed. The
deslimed product was passed through a Frantz isodynamic separator
to remove hematite. The remaining product was passed through
heavy liquids having sp. gravities of 2.96 and 2.78 and 2.68.
Each fraction was weighed, analysed by X-ray diffractometer to
obtain semi-quantitative mineral analyses, then assayed for gold.
The highest gold concentration occur in slimes which account for

77.5% of the gold and lowest amount in the quartz rich fraction. A plot of clay percentages against gold values show that there is a linear correlation between clay and gold content. More than 90% of the gold was extracted by cyanidation.

5.7 Case VII - Carlin Gold

Gold in some of the Carlin deposits of Nevada occurs in pyrite [27]. There is no apparent correlation between gold and pyrite content and a variable Au extraction was obtained. Mineralogical studies failed to reveal much gold but two forms of pyrite were recognized: a crystalline to massive pyrite and a fine spheroidal pyrite that in some cases is nearly framboidal. Grain counts were performed to determine the distribution of different forms of pyrite. The samples were then treated by cyanidation and the % of gold extracted Vs % spheroidal pyrite were plotted. The results show a linear relationship which indicates that submicroscopic gold occurs in the very fine-grained spheroidal pyrite and is readily extracted by leaching.

The gold in crystalline and massive pyrite is coarser grained, however, is completely enclosed in the pyrite and is not leachable. A chlorination-cyanidation technique was developed by Newmont Mines Limited to dissolve the crystalline pyrite and recover the gold. Tests on samples treated by this technique show that there is a direct relationship between the pyrite removed and the extraction of gold.

5.8 Case VIII - Selective Flotation of Sulphides from Oxidized Base Metal Ores

This is a major problem that arises whenever selective flotation of base metal ore from an oxidized zone is attempted. In normal flotation of base metal ores copper and lead concentrates are floated first, then sphalerite is floated from the Cu-Pb tailings. When the ore contains more than 0.01 wt % covellite or chalcocite the sphalerite floats along with the chalcopyrite. On one ore that contained about 5 wt % Zn, 3 wt % Pb, and 0.5 wt % Cu, a selective float could not be obtained when about 1% of the copper (.005 wt % Cu) was covellite plus chalcocite. The ore was treated with Na cyanide and the sphalerite was effectively depressed.

5.9 Case IX - Silica and Goethite in Iron Ore

Pellets that are produced from iron concentrates have to meet several specifications; the silica content must be low and the pellets must have a high strength. The silica content is usually due to quartz in the iron concentrate. Sometimes the quartz in the concentrate is as free particles, and can be rejected with proper plant performance, whereas at other times it is unliberated and regrinding is required. Periodic monitoring of the concentrate with an ore microscope is required to maintain specifications.

The problem of pellet strength is often related to the amount of goethite and/or biotite in the concentrate. The minerals contain water as part of their composition. This water is released during sintering of the pellet and causes the pellets to crack. Periodic monitoring of the concentrate with an ore microscope is necessary to establish that the minerals are rejected during ore dressing.

5.10 Case X - Oolitic Iron Ores

A very large deposit of oolitic iron ores occurs in the Peace River District in Alberta, Canada. It contains about 37 wt % iron. Microscopical, microprobe, SEM, X-ray diffraction, infra-red and Mossbauer analysis showed that the ore is composed of sphere-like bodies in a matrix of ferruginous opal. The sphere-like bodies are composed of a goethite core surrounded by nontromite interlayered with goethite. The goethite contains about 62 wt % Fe, nontronite 39 wt % Fe, and ferruginous opal 24 wt % Fe. Image analysis studies determined that the ore contains about 35 wt % goethite, 40 wt % nontromite and 25 wt % ferruginous opal. The best concentrate that could be obtained by ore dressing would contain about 48 wt % Fe, and the best recovery would be 60 to 65 wt % of the iron [28]. These results are so poor that pyrometallurgical tests were successfully conducted to recover the iron directly. More than 90% of the iron would be recovered by this technique but the energy costs are high. The deposit has not been brought into production but the technology for recovering the metal from the ore has been developed.

5.11 Case XI - Titanium Ore

Ore dressing tests were conducted on a titanium ore from Quebec, Canada. The ore consists of ilmenite, magnetite, hematite and silicate minerals. The minerals are coarse grained and can be separated with a magnetic separator. The best ilmenite concentrate contained 39 wt % titanium, even though stoichiometric ilmenite contains 52 wt % Ti. Image analysis studies determined that the ilmenite contained 20 wt % hematite as lamellae 0.1 to 3 μm wide. Therefore the ilmenite can contain a maximum of 42 wt % Ti [29].

6. EQUIPMENT FOR PERFORMING MINERALOGICAL ANALYSIS

6.1 Microscopes

A. <u>Binocular</u>. used for observing hand specimens and mill products at low magnification.

B. <u>Transmission and Reflecting</u>. used for observing thin sections in transmitted light and polished sections in reflected light. The minerals are identified on sight. Particles as small as 1 μm in diameter can be observed, but to enable identification most particles must be larger.

6.2 X-ray Diffraction

Individual particles 35 μm or larger are dug out of polished sections and mounted on glass fibres. Deby-Scherrer x-ray powder patterns are obtained and used to identify the minerals. A Guiner camera is used for clay minerals but a few milligrams of the sample is required.

6.3 X-ray Diffractometer

About 1 gram of crushed powder is used. Tracings of the powder patterns are obtained. Minerals are identified using peak positions, and the peak intensities on the tracings are proportional to mineral quantities.

6.4 Electron Microprobe

Polished sections are irradiated with an X-ray beam and compositions of grains larger than 3 μm can be determined.

6.5 Scanning Electron Microscope Equipped with Energy Dispersive X-ray Analyser (EDXA)

An image of the polished section is obtained. The most useful image is electron backscatter where the grey level is proportional to the Z number of the mineral. The approximate composition of each phase can be determined with the EDXA thus the mineral can be identified.

6.6 Infra-Red Spectroscopy

Analysis is performed by mixing the crushed material with KI and pressing it into a thin wafer. The wafer is mounted and either an infra-red transmission or absorption spectrum is obtained.

Peaks at different positions indicate the presence of different molecules such as SiO_2, H_2O, OH, SO_2, CO_2 etc. The intensities of the peaks are semi-quantitatively proportional to quantities of the mineral. Infra-red is particularly useful for minerals that have a poor crystal structure and have the above molecules, i.e., for anglesite, goethite, clay minerals, etc. The technique has been used successfully for determining quantities of quartz in mine dust, and for identifying nontronite (a clay mineral) and ferrugnious opal (a non-crystalline mineral).

6.7 DTA - TGA

Is used to measure endothermic and exothermic reactions and is useful when combined with TGA analysis of non-crystalline minerals that lose weight upon being heated. The method of analysis is to take a small amount of powder and heat to say 1000°C. Thermocouples measure the endothermic and exothermic reactions that are distinctive for the mineral and TGA analysis measures the weight loss that occurred with each reaction. The technique is particularly useful when the minerals do not have well developed crystal structures, such as non-crystalline goethite, and have much OH, CO_2, P_2O_5 or other complexes that are driven off as the mineral is heated.

6.8 Mossbauer Spectroscopy

Absorption spectra give distinctive peaks for ferric and ferrous ions. The peak positions are distinct for each mineral and the intensities of the peaks are roughly proportional to quantities.

Mineral identification methods that do not require sophisicated instrumentation are staining, fluorescence, contact printing, radiography, and others. Most of these techniques have been replaced by the newly developed instruments. The most easily applicable and readily available stain is alizarin red for carbonate minerals, described by Warne in about 1961. The alizarin red solution stains calcite blood red, dolomite blue, and siderite brown. Another staining technique is the procedure for staining polished sections. There are six standard reagents. A drop of each reagent is applied to the surface of a polished section using a platinum wire. The drop is observed through the microscope for about 30 seconds to 1 minute. The reagent is then washed with distilled water and the spot is again observed. This procedure was used with great success before the widespread use of X-ray diffraction and microprobe analysis.

Fluorescence in ultra-violet light is used to identify some minerals such as scheelite, autunite, some calcites and other minerals.

The printing of radioactive minerals is used to some extent. A film is placed on the mineral for a controlled period of time, and then developed. The imprint of where alpha particles bombarded the film can usually be identified. By counting the alpha-tracks per cm^2 per hour the variety of radioactve mineral can be identified.

6.9 SEM-Based Automatic Image Analysis System:

This is the most useful equipment in applied mineralogy, therefore the system at CANMET is described in some detail, as an example. The CANMET system consists of a Jeol 733 electron microscope (SEM) equipped with 2 wavelength spectrometers and a Tracor Northern energy dispersive X-ray analyser (EDXA) interfaced with a Kontron image analyser (IA) that has 16 memories. The automatic image analysis system (AIA) operates by producing an image with the scanning electron microscope. The backscatter electron image is the most useful image because phases having different compositions display different grey levels. The image is taken into memory 1 of the image analyser where a series of filters are applied so that each grey level can be readily discriminated. Each filtering step moves the image into a different memory bin with the final image being stored in memory 2. Each grey phase is then detected separately, one per memory and an outline is produced for each phase. A composite of the particle outlines is then produced and the composite is used as a "psuedo-live-time image" for controlling the position of the electron beam to scan the selected features in the field. The KONTRON image analyser isolates each particle and each feature within the particle is scanned with the SEM. The X-ray spectrum for each feature is collected with the EDXA and sorted into elements, peak intensities being proportional to quantities of elements. The sorted spectrum is sent to the KONTRON IA where it is compared to a mineralogy file and each feature is identified. Object specific data such as sizes and shapes of the particles and of inclusions in the particles are collected by the Kontron IA. Conditions for the object specific subscans are defined interactively but the analysis is performed automatically.

Another major use of the image analyser at CANMET is in the location and subsequent analysis of minerals which occur in very low concentrations in other materials, such as tailings. In these cases there may be only a single grain of a valuable mineral, e.g. native gold, in many thousands of grains of gangue minerals. One approach to this problem is to use the linear traverse technique while monitoring the output of the wavelength spectrometers which are set for specific elements of interest. When the elements sought are detected, then the grain of interest is positioned under the beam by the image analyser, which controls the stage, and the measurement frame is erected on the grain for stereological measurements. Subsequently, the grain may be quantitatively analysed using the two wavelength spectrometers at a later time.

7. METHODS OF MINERAL IDENTIFICATION

A detail of discussion on methods of mineral identification is beyond the scope of this paper, but methods of mineral identification are listed below.

1. Optically – binocular microscope
              – transmission microscope
              – reflecting microscope

2. X-ray powder diffraction

3. Electron microprobe

4. SEM equipped with EDXA

5. Other methods – infra-red analysis
                  – ultra violet analysis
                  – Mossbauer analysis
                  – DTA and TGA analysis
                  – radiography
                  – staining

8. METHODS OF DETERMINING MINERAL COMPOSITIONS

1. Microprobe

2. SEM equipped with EDXA

3. X-ray powder diffraction

4. Chemical analysis of mineral concentrates

9. METHODS DETERMINING MINERAL QUANTITIES

There is no simple mineralogical technique to quickly and accurately determine mineral quantities. Some techniques such as X-ray diffractometer analysis give semi-quantitative analysis quickly, and others, such as image analysis can give good quantitative data, but this technique is new and not yet well-developed. The standard direct analysis is by separating the minerals into fractions, weighing each fraction, and determining the mineral content of each fraction by X-ray diffractometer, image analysis, etc. Another method of determining mineral quantities is performing a semi-quantitative mineral analysis, determining the chemical composition of the sample, then calculating the mineral content.

The minerals can be separated into into fractions using:

1) Heavy liquids
2) Elutriating tube
3) Frantz isodynamic magnetic separator
4) Electrostatic separator
5) Superpanner
6) Flotation

The separated fractons can be analysed with

1) X-ray diffractometer
2) Estimation under a microscope
3) Grain counting
4) Image analysis

10. METHODS OF DETERMINING MINERAL LIBERATION AND SIZE DISTRIBUTIONS

1) Measuring chord length under an optical microscope

2) Image analysis

11. PRESENTATION OF DATA

When mineralogical data are collected they should be presented in easily understandable formats. Data that would be obtained from uncrushed ore fragments are mineral identities, modal analyses, and size distribution of the minerals of interest. Mineral identities and modal analyses would be presented in standard tables. Size distribution data for each

mineral of interest can be presented in either tables (Table 1) or
in graphical presentations that give a quick visual perspective of
the size distributions (Figs. 5 and 6). The graphical
presentation of cumulative percent vs size, plotted on semi-log
graph paper is similar to the method that mineral dressing
engineers use for presenting screen analysis data.

Table 1. Size distributions of minerals in uncrushed sample

Size range µm	Sphalerite		Galena		Chalcopyrite	
	Vol. %	No.*	Vol. %	No.*	Vol. %	No.*
0-4.6	21	4549	36	4650	27	4550
4.6-6.5	6	227	7	158	8	236
6.5-9.2	5	89	9	96	6	84
9.2-13	7	62	9	47	9	62
13-18.5	7	32	10	28	10	37
18.5-26	9	20	9	12	10	18
26-37	8	9	8	6	10	9
37-52	10	6	6	2	10	4
52-74	8	2	3	1	4	1
74-104	7	1	2	1	4	1
104-147	4	1	1	1	1	1
147-208	5	1	0		1	1
+208	3	1	0		0	0
Total	100	5001	100	5001	100	5004

*number of particles

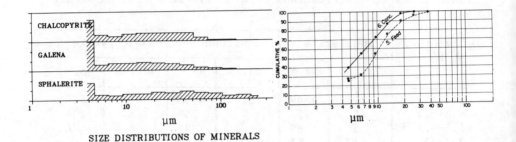

SIZE DISTRIBUTIONS OF MINERALS

Figure 5 - Size distributions of sphalerite, galena and chalcopyrite (data of Table 1), plotted as % in each size range of the Tyler Series. The vertical scale is 0-37%.

Figure 6 - An example of size distribution of sphalerite plotted semi-log graph paper as cumulative % vs size in micrometres

11.1   Presentation of Data for Mill Products

Mill products are obtained from laboratory bench tests, pilot plant tests, and from a plant. In some instances the mill products can be analysed directly without screening into fractions, but in other instances, particularly if the mill product is coarse grained, it has to be screened. Data that would be obtained by image analysis from polished section of either the screened fractions or of the mill products are:

1) Modal analysis,

2) Proportions of minerals of interest occurring as free particles and as unliberated grains,

3) Proporation of multi-phase grain that is occupied by mineral of interest,

4) Size distributions of the free particles, of the host
particles, and of the inclusions of the mineral of interest in
the host particles.

When the analysis is performed on screened fractions, size
analysis is not required for the free mineral particles nor for
the host particles. However, the weight per cent, and often the
assays of each sized fraction (Table 2) are needed so that the
mineral data for the screened fractions can be recast to mineral
data for the mill product.

Table 2. Assays, element distributions, and weights of
sized fractions

Fraction (μm)	Wt. %	Assay %			Distribution %		
		Zn	Cu	Pb	Zn	Cu	Pb
−10	6.91	10.54	0.74	6.44	7.99	10.60	13.99
10–15	9.29	9.17	0.55	6.10	11.70	10.56	17.82
15–25	7.59	8.61	0.53	3.14	8.96	8.28	7.50
25–37	7.71	7.22	0.47	2.30	7.63	2.59	5.70
37–52	4.36	7.10	0.48	2.35	4.18	4.24	3.24
52–74	4.41	6.90	0.46	2.38	4.11	4.29	3.32
74–104	4.73	6.66	0.46	2.41	4.22	4.60	3.56
104–147	3.85	6.64	0.45	2.42	3.45	3.66	2.93
147–208	5.81	6.62	0.44	2.48	5.20	5.40	4.53
208–294	7.36	6.60	0.44	2.51	6.40	6.62	5.70
294–416	8.11	6.57	0.44	2.56	7.02	7.29	6.38
416–588	7.14	6.53	0.44	2.60	6.15	6.42	5.72
588–832	22.73	7.29	0.48	3.18	20.99	20.45	19.61
Combined Product	100.00	7.52	0.50	3.29	100.00	100.00	100.00

11.2 Modal Analysis

Modal analyses can be determined by point counting, by
measuring linear intercepts, and by measuring the areas of the
grain surfaces exposed in polished sections. The modal analysis
for each size fraction and the distribution of minerals among the
various fractions is reported in standard tables (Table 3) and
calculated for the combined mill product.

Table 3. Modal analysis

Fraction μm	% Mineral					Distribution of minerals				
	sp	cp	ga	py	sil	sp	cp	ga	py	sil
-10	17.6	2.2	7.5	48.2	24.5	1.22	0.15	0.52	3.33	1.69
10-15	15.3	1.6	7.1	50.1	26.4	1.42	0.15	0.66	4.65	2.45
15-25	14.4	1.6	3.7	52.0	28.6	1.09	0.12	0.28	3.95	2.17
25-37	12.0	1.4	2.7	56.3	27.6	0.93	0.11	0.21	4.34	2.13
37-52	11.8	1.4	2.7	56.5	27.6	0.51	0.06	0.12	2.46	1.20
52-74	11.5	1.4	2.8	57.5	26.8	0.51	0.06	0.12	2.54	1.18
74-104	11.1	1.4	2.8	58.0	26.7	0.53	0.07	0.13	2.74	1.26
104-147	11.1	1.3	2.8	58.5	26.3	0.43	0.05	0.11	2.25	1.01
147-208	11.0	1.3	2.9	58.1	26.7	0.64	0.08	0.17	3.38	1.55
208-294	11.0	1.3	2.9	58.3	26.5	0.81	0.10	0.21	4.29	1.95
294-416	11.0	1.3	3.0	58.2	26.5	0.89	0.11	0.24	4.72	2.15
416-588	10.9	1.3	3.0	58.0	26.8	0.78	0.09	0.21	4.14	1.91
+588	12.2	1.4	3.7	56.0	26.7	2.77	0.32	0.84	12.70	6.04
Combined Product	12.5	1.5	3.8	55.5	26.7	12.53	1.47	3.82	55.49	26.69

sp=sphalerite, cp=chalcopyrite, ga=galena, py=pyrite, sil=silicate minerals

11.3 Proportion of Mineral as Free Particles or Unliberated Mineral Grains

The proportions of minerals that are present as free particles and as unliberated mineral grains (Fig. 7) must be determined to establish whether the ore has been ground sufficiently to liberate the minerals, and also to evaluate the performance of a concentrator. The liberation of the minerals of interest in each screened fraction can be determined with an image analyser by measuring either the surface area or linear intercepts of the free particles and unliberated grains of the mineral in polished sections and calculating the percentage that is free. These data are not absolute because unliberated grains mounted in polished sections might be sliced so that they appear free [10,11] hence the observed amount of free particles will be slightly higher than the true value. Nevertheless, for practical purposes the data are close enough to the true values and should be presented in standard tables (Table 4).

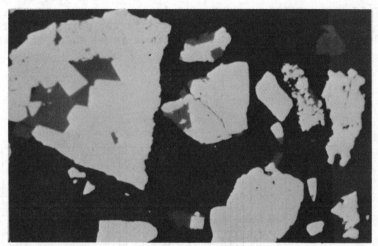

Figure 7 - Photomicrograph showing free and unliberated sphalerite
grains (grey) in a -150 mesh (104 μm) fraction

Table 4. Mineral liberation

Fraction μm	% Free	Sphalerite, %
	In fraction	Contribution to mill product
-10	98	1.20
10-15	87	1.24
15-25	73	0.80
25-37	60	0.56
37-52	42	0.21
52-74	36	0.18
74-104	28	0.15
104-147	18	0.08
147-208	6	0.04
208-294	0	0.0
294-416	0	0.0
416-588	0	0.0
+589	0	0.0
Total		4.46
% free in mill product		4.46/12.53=35.6

To establish the minimum and optimum grinds for a mineral in the ore, a series of sized fractions is analysed to determine the per cent of the mineral as free particles. The data are plotted on semi-log graph paper as per cent free vs. grains size for each fraction (Fig. 8). The minimum grind is the point where the curve begins to steepen (100 µm), and optimum grind where it flattens (15 µm).

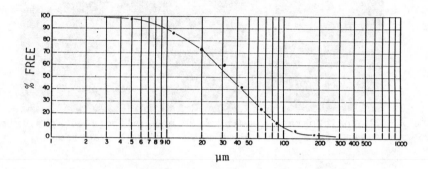

Figure 8 - Percent free sphalerite in each sized fraction. The vertical scale is % free and the points are plotted at the midpoint of each sized fraction.

The contribution of the free grains to each fraction in the total sample can be calculated from the data for each screened fraction by multiplying the mineral distribution of Table 3 x percent free (Table 4).

Plant performance can be evaluated by determining the behaviour of free mineral grains in the concentrator or a circuit. This involves:

1) collecting a complete suite of mill products from the concentrator or circuit;

2) assaying these products for elements of interest;

3) performing a materials balance calculation (with a MATBAL program) [12], using the assay data, to calculate the weight per cent of each mill product relative to the feed for the concentrator or circuit (Table 5);

4) performing a screen analysis on each mill product and
    weighing each fraction;

5) performing a modal analysis on each screened fraction and
    calculating the modal analysis for the mill product (see
    example in Table 3);

6) determining per cent of mineral of interest that is present
    as free particles in each fraction and calculating the
    quantity (see example in Table 4) and size distribution of
    free particles in each mill product;

7) recalculating the quantity (Table 6) and the sizes of free
    particles and of unliberated grains in each mill product
    relative to free particles in the feed.

8) plotting the data of the previous step as curves of per cent
    free vs. size (Fig. 9). To prepare these plots one must
    have data for each systematic size interval; the most
    convenient set of interval is the Tyler series. On this
    plot the top curve represents the quantity of free particles
    in the feed, the bottom curve, the quantity in the
    concentrate, and the space between the curves represents the
    amount vs. grain size lost to the tailings and other
    products. For more detail the amounts lost to tailings and
    other products are also plotted.

Good mill performance is indicated if the recovery of free
mineral particles in the concentrate is about 90 to 95% (calculated
from Table 6). The loss of either fine-grained or coarse-grained
free mineral particles can be readily seen from Fig. 9 and
appropriate corrective steps can be taken.

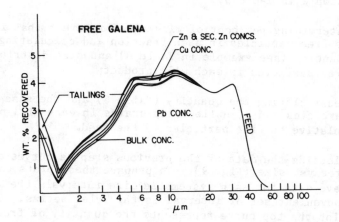

Figure 9 – Frequency diagram showing quantity of free galena
particles in each size range of the Tyler series in the
feed (top curve) and in the lead concentrate (bottom
curve). The spaces between the curves represent the
quantities of free galena lost to the tailings and other
products.

Table 5. Distribution of products, determined by MATBAL

Product	Wt. %	Assay			Distribution		
		Zn	Pb	Cu	Zn	Pb	Cu
Cu conc.	0.42	2.34	15.42	22.00	0.11	1.47	30.8
Pb conc.	8.00	8.81	34.96	0.73	7.20	64.91	14.2
Zn conc.	17.80	51.63	3.51	0.22	84.56	12.56	16.1
Tailings	73.78	1.35	1.16	0.12	8.13	21.06	38.9
Feed	100.00	9.47	4.49	0.30	100.00	100.00	100.0

Table 6.   Mineral liberation

Product	% Sphalerite		% Galena		% Chalcopyrite	
	Free	Unliberated	Free	Unliberated	Free	Unliberated
Cu conc.	0.05	0.10	1.3	0.2	27.5	3.3
Pb conc.	4.00	3.2	32.6	32.4	9.3	4.9
Zn conc.	60.90	23.6	1.4	10.9	8.9	7.2
Tailings	1.60	6.6	1.7	19.5	10.3	28.6
Feed	66.5	33.5	37.0	63.0	56.0	44.0

11.4   Percent Inclusions of Unliberated Valuable Minerals

The unliberated mineral grains are studied to assess the
maximum possible recovery from an ore. An unliberated grain is
defined in this report as the grain of the desired mineral in a
particle that contains two or more minerals (Fig. 7). The
characteristics of the particle must be assessed to determine
whether the particle would tend to report to either the concentrate
or tailings, or whether it should be reground. The approximate
percentage of the desired mineral in the particle can be determined
by measuring the surface area of each mineral grain and of each
particle, and by calculating, by computer, the percentage of
mineral in each particle. The data can be presented as bar graphs
with a bar for each category of particle, i.e. particles that
contain the following quantities of desired mineral in vol. %
0-10, 10-20, 20-30, 30-40, 40-50, 50-60, 60-70, 70-80, 80-90,
90-100% (Fig. 10).The category of 90-100% mineral in the particle
would be classed as free particles. The height of each bar
represents the percentage of the mineral that is present in a
specific category of particle. The data for each fraction can be
recalculated to data for each mill product, using the weight of
each fraction, and replotted to show the contribution of each
fraction to the mill product. Size distributions of the particles
and of the mineral inclusions in particles of each category would
be determined with the image analyser and plotted as histograms
or scatter diagrams (Fig. 11).

32

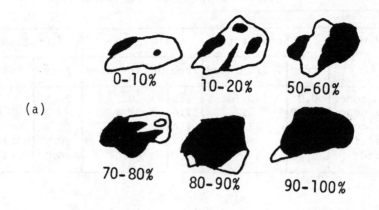

(a)

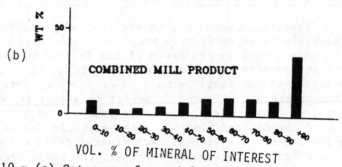

(b)

Figure 10 - (a) Category of particle, black is the mineral of interest. (b) Bar graph showing percent of mineral that is in each category of particle.

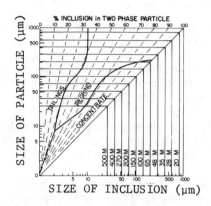

Figure 11 – Diagram for plotting size of inclusion vs size of its
host particle. Schematic domains for sphalerite-
bearing particles in a high grade zinc concentrate, and
in tailings are shown. Particles with intermediate
quantities of sphalerite would need to be reground.
The numbers at the top of the diagram and the dashed
lines indicate percentages of wanted mineral in a
two-phase particle. The solid lines give mesh sizes.

REFERENCES

1. Petruk, W. Applied Mineralogy in Ore Dressing of Volcanogenic
    Massive Sulphide Ores, in ICAM 84, TMS/AIME, New York, 1985,
    Park, W., et al, (eds), in press.

2. Shapiro, N., Mallio, W.J. and Park, W.C. Process Mineralogy
    in Ore Deposits Development, in Process Mineralogy, TMS/AIME,
    New York, 1981, Hausen, D.M. and Park, W. (eds), pp. 25.

3. Petruk, W. and Schnarr, J.R. An Evaluation of the Recovery of
    Free and Unliberated Mineral Grains, Metals and Trace Elements
    in the Concentrator of Brunswick Mining and Smelting Corp.
    Ltd., CIM Bull. 78 (833), 1981, pp. 132.

4. Gaudin, A.M. "Principles of Mineral Dressing", McGraw-Hill
    Inc., New York, 1939, pp. 70.

5. Melvik, T. An Empirical Model to Establish the Liberation
    Properties of Minerals, Proc. 14th Intl. Min. Processing
    Cong., Toronto, CIM 1982, Preprint No. VIII-2.

6. King, R.P. The Prediction of Mineral Liberation from
    Mineralogical Texture, (14th Intl. Min. Processing Cong.,
    Toronto, CIM 1982), Preprint No. VIII-1.

7. Klimpel, R.R. Some Practical Approaches to Analyzing
    Liberation from a Binary System, in Process Mineralogy III,
    SME/AIME, New York, 1984, Petruk, W. (ed), pp. 65.

8. Petruk, W. The Application of Quantitative Mineralogy
    Analysis of Ores to Ore Dressing, CIM Bull. 69 (767), 1976,
    pp. 146.

9. Laguitton, D., Flament, F., Wilson, J. and Everell, D. SPOC
    Manual Chapter 3A: Material Balance Computation for Process
    Evaluation and Modelling: BILMAT Computer Program, Division
    Report MRP/MSL 83-82 (IR), CANMET, Energy, Mines and
    Resources, Canada, 1983.

10. Petruk, W. Correlation Between Grain Sizes in Polished
    Section with Sieving Data and Investigation of Mineral
    Liberation Measurements from Polished Sections, Inst. Mining
    and Met., Trans. 87, 1978, C272.

11. Stemerowicz, A. and Sirois, L. Concentration of Complex
    Sulphide Ores, CIM Bull., 1984, in press.

12. Hagni, R.D. The Application of Reflected and Luminescence Microscopy and the Auger Microprobe to Beneficiation Problems, 14th Intl. Min. Processing Cong., Toronto, CIM, 1982, Preprint No. VIII-7.

13. Amstutz, G.C. How Microscopy Can Increase Recovery in Your Milling Circuit, Mining World, Dec. 1962, pp. 19.

14. Carson, D.J.T. Geological and Mineralogical Controls on the Metallurgy of the Grum Massive Sulphide Deposit, Yukon Territory, CIM Bull. 71 (791), 1978), pp. 106 (abstract).

15. Jones, M.P. Designing an X-ray Image Analyser for Measuring Mineralogical Data, 14th Intl. Min. Processing Cong., Toronto, CIM 1982, Preprint No. VIII-4.

16. Rixom, P.M. and Kostic, N. The Application of Image Analysis to Determine the Flotation Characteristics of Copper and Iron Sulphides from the Roan Antelope Deposit in Zambia, 14th Intl. Min. Processing Cong., Toronto, CIM 1982, Preprint No. VIII-5.

17. Miller, P.R., Reid, A.F. and Zuiderwyk, M.A. QEM-SEM Image Analysis in the Determination of Modal Assays, Mineral Associations and Mineral Liberation, 14th Intl. Min. Processing Cong., Toronto, CIM 1982, Preprint No. VIII-3.

18. Chen, T.T. and Petruk, W. Mineralogy and Characteristics that Affect Recoveries of Metals and Trace Elements from the Ore at Heath Steele Mines, New Brunswick, CIM Bull. 73 (823), 1980, pp. 167.

19. McLean, D.C. Upgrading of Copper Concentrates by Chalcocite Derimming of Pyrite with Cyanide, in Process Mineralogy III, SME/AIME, New York, 1984, Petruk, W. (ed.), pp. 3.

20. Anon, Westmin Resources: Profile of Progress, Western Miner, May 1983, pp. 10.

21. Petruk, W. Impurities in the Lead Concentrate of Brunswick Mining and Smelting and their Behaviour in the Lead Furnace, CIM Bull., 1985, (in press).

22. Boorman, R.S., Petruk, W. and Gilders, R. Silver Carriers in Concentrate and Tailings from Brunswick Mining and Smelting Corporation Limited, in Process Mineralogy III, SAME/AIME, New York, 1984, Petruk, W. (ed.), pp. 33.

36

23. Hagni, R.D.   Applications of the Ore Microscope to Ore
    Dressing Problems in the Viburnum Trend, South-East Missouri,
    USA, in Proc. ICAM 81, proceedings, sp. Pub. No. 7, Geol.
    Soc. of South Africa, Randburg, 1983, DeVilliers, J.P.R. and
    Cawthorn, P.A. (eds.), pp. 209.

24. Haagensen, R.B. and Martinez, E.   Practical Applications of
    Microscopy in Mineral Processing, in Process Mineralogy,
    TMS/AIME, New York, 1981, Hausen, D.M. and Park, W. (eds.),
    pp. 213.

25. Sizgoric, M.   Alteration of Nickel Sulphide Ores and its
    Effect on their Flotation, in Process Mineralogy, TMS/AIME,
    New York, 1981, Hausen, D.M. and Park, W. (eds.) pp. 225.

26. Ahlrichs, J.W.   Application of Quantitative Mineralogy for
    Solving Metallurgical Problems, in Process Mineralogy,
    TMS/AIME, New York, 1981, Hausen, D.M. and Park, W. (eds.),
    pp. 271.

27. Hausen, D.M.   Process Mineralogy of Auriferous Pyritic Ores at
    Carlin, Nevada, in Process Mineralogy, TMS/AIME, New York,
    1981, Hausen, D.M. and Park, W. (eds.), pp. 271.

28. Petruk, W., Klymowsky, I.B. and Hayslip, G.O.   Mineralogical
    Characteristics and Beneficiation of an Oolitic Iron Ore from
    Peace River District, Alberta, CIM Bull. 70 (786), 1977, pp.
    122.

29. Petruk, W.   Mineralogical Examination of a Titanium Ore from
    the Gauthier Property in Quebec, Division Report IR-72-63,
    CANMET, Energy, Mines and Resources, Canada, 1972.

CHEMISTRY OF SULPHIDE MINERAL FLOTATION

Raşit Tolun

Professor of Chemical Engineering Department, Faculty of
Chemistry - Metallurgy, İstanbul Technical University,
Maslak - İstanbul, Turkey

ABSTRACT

Chemistry of sulphide mineral flotation by xanthates has been
reviewed and discussed with respect to hydrophibization by collector coating and depressing effects of hydroxyl, cyanide and sulphide
ions. The collection mechanism in Galena-Oxygen-Xanthate Flotation
System has been elucidated electro-chemically by the use of electrode potentials. It is concluded that there is a limiting range of
Eh and pH where flotation is possible. Finally, silver sulphide
electrode was utilized to control the activity of xanthate ion during flotation of oxidized lead minerals after sulphidization. The
continuous control of sulphide and xanthate ions has been found practical to achieve reproducible results during flotation of an oxidized
lead-zinc ore containing cerussite, smithsonite and hydrozincite.

1. INTRODUCTION

Minerals may be classified, according to their behaviour in the
flotation process as the following [1] :

1. Non-Polar, Non-Metallic Minerals (graphite, sulphur, coal
   and talc).

2. Sulphides of Heavy Metals and Native Metals (galena, sphalerite, chalcopyrite, antimonite etc.).

3. Oxidized Minerals of Non-Ferrous Metals (cerussite, smithsonite, malachite etc.).

4. Polar Salt-Type Minerals which Contain Cations of Calcium, Magnesium, Barium and Strontium.
   (calcite, fluorite, apatite, barite, magnesite etc.).

5. Oxides, Silicates and Aluminosilicates.
   (silica, feldspars, hematite, pyrolusite etc.).

6. Soluble Salts of Alkali and Alkaline Earth Metals. (sylvite, langbeinite, gypsum, colemanite etc.).

The ease of floatability of these minerals decreases as the ionic character of the chemical bonding increases. The gradation between covalent and ionic bonding is simply presented by the illustration below [2, 11] .

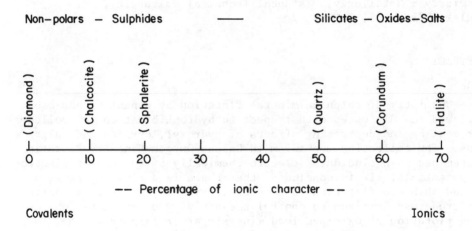

Figure 1. Gradation between ionic and covalent bonding

Beside being of poor ionic character, most of the sulphide minerals differ from others by showing electrical conductivity, but compared with metals they are extremely poor conductors (semimetallics), differing from them by several order of magnitute. Some of them display the properties of allowing passage of current in one direction only (semiconductors). In fact this unidirectional valvelike quality of galena was utilized in early, crystal radio sets [3] .

The most important characteristic of sulphide minerals is their instability in atmospheric conditions. They transform slowly into stable oxidized minerals, e.g. $PbS \longrightarrow PbSO_4$, $ZnS \longrightarrow ZnCO_3$, etc.

Considerable changes on the surface of sulphide minerals occur

during their oxidation. Numerous data relate the kinetics of the
oxidation of sulphides to the formation of thick oxide layers and
finally to the complete transformation of sulphides into oxides.
In the following series, some of the sulphides are listed showing
a diminishing tendency toward oxidation :
Arsenopyrite > Pyrite > Chalcopyrite > Sphalerite > Galena > Chalcocite.

Sulphides of old geological formations are less readily oxidized than those of later formations. Two adjoining solid phases are
much more readily oxidized than a single substance. Thus, galena,
sphalerite and covellite oxidize from eight to twenty times faster
in the presence of pyrite.

The oxidation of sulphides in flotation systems, where changes
in the surface layers of molecular dimensions play an important role
is still of a preliminary nature. An increase of the oxidation
species has been observed with an increase in pH, with the concentration of dissolved oxygen and with the time of retention of minerals in the pulp [4] .

1.1 Sulphide Minerals-Water-Oxygen System

In order to understand clearly, the oxidizing effect of oxygen
on sulphides in aqueous solutions, electrochemical equilibria diagrams should be compared [5] .

Dissolved oxygen of air in water, may act as an oxidizing agent,
by reducing itself into water.

$$O_2 + 4H^+ + 4e^- = 2H_2O \quad , \quad E^o = 1.23 \text{ Volt}$$

As a result of this reaction conducting (or semiconducting)
inert surfaces should acquire an electrochemical equilibrium potential, depending on the partial pressure of oxygen and hydrogen ion
concentration :

$$E = 1.23 - \frac{0.059}{4} \log \frac{1}{(H^+)^4 \ P_{O_2}} \qquad \text{or}$$

$$E = 1.23 - 0.059 \text{ pH} + \frac{0.059}{4} \text{ Log } P_{O_2}$$

In Figure 2, the line b indicates the change of the electrochemical potential versus pH when the pressure of oxygen is one
atmosphere

$$E = 1.23 - 0.059 \text{ pH}$$

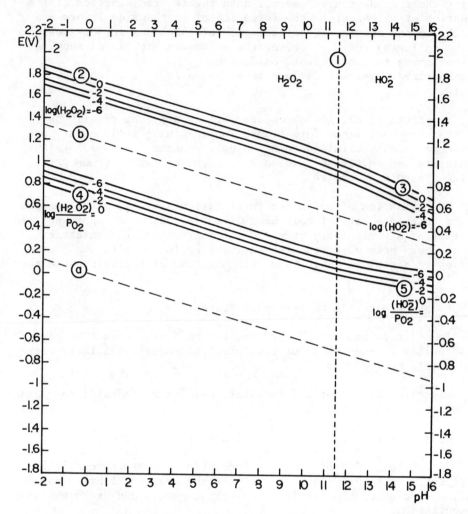

Figure 2. Equilibrium diagrams of E-pH of $H_2O - H_2O_2$
system at $25^{\circ}C$ 5 .

In practice, the potentials reach lower values (lines 4 and 5),
indicating that the first step of the reduction of oxygen is the
potential determining reaction [7] .

$O_2 + 2H^+ + 2e^- = H_2O_2$, $E^o = 0.68$ Volt

$E = 0.68 - 0.059 \text{ pH} + 0.030 \log P_{O_2} - 0.030 \log (H_2O_2)$

and at high pH's, ($pK_a = 11.6$ for $H_2O_2 = HO_2^- + H^+$)

$O_2 + H^+ + 2e^- = HO_2^-$, $E^o = 0.34$ volt

$E = 0.34 - 0.030 \text{ pH} + 0.030 \, P_{O_2} - 0.030 \log (HO_2^-)$

H_2O_2 produced by this reaction, is decomposed.

$2H_2O_2 \longrightarrow 2H_2O + O_2$

Thus, its concentration does not exceed 10^{-6} M. Assuming that the partial pressure of oxygen is about 0.2 atm., the equilibrium lines of 4 and 5 corresponding to $\log (H_2O_2 / P_{O_2}) = 5.3$, should be placed between the lines indicated by −6 and −4 in Figure 2.

The line a of the figure represents the lower limit of water stability. Below that line, water is reduced with hydrogen evolution :

$2H_2O + 2e^- = H_2 + 2OH^-$ or $2H^+ - 2e^- = H_2$, $E = -0.059 \text{ pH}$

Hence any reducing matter, e.g. sulphide minerals, having potentials lower than $O_2 - H_2O_2$ equilibrium lines 4 and 5 in aqueous solutions, are stable only when they are isolated from the atmospheric oxygen.

Figure 3. summarizes the positions of some natural environments.

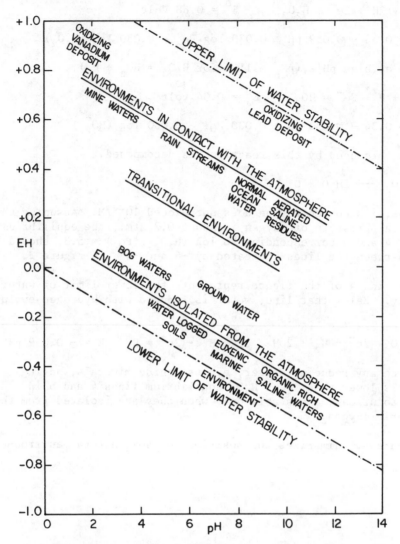

Figure 3. Approximate position of some natural environ-
        ments as characterized by Eh   and   pH [6] .

Hence, depending on the pH,oxygen content of the solution and
of the presence of the common and other potential determining ions,
sulphide minerals acquire different electrochemical potentials.
Furthermore, the non-stochiometry and the previous history influence
these potentials.

It was observed that the reversible potentials established
by the metal sulphides in contact with solutions containing the
corresponding cations have, however, been found to be more noble

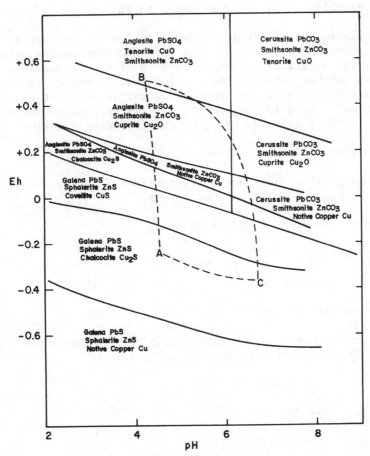

Figure 4. Composite diagram of stability of metal sulfides and oxidation products at 25°C and 1 atmosphere total pressure in the presence of total
dissolved carbonate = 10^{-15}, total dissolved
sulfur = 10^{-1} [6] .

Figure 4, is an interesting illustration, indicating the regions
of stabilities of lead, zinc and copper sulphides and their oxidation
products.

2. MODULATION OF COLLECTOR COATING

2.1 Hydrophobization of Sulphides by Xanthate Coating

In this study, potassium ethyl xanthate has been taken as the
collector representing dithio compounds; because, it is the most
used and the best studied amongst all of them.

than those established by the respective parent metals [9] .

The oxidation of a sulphide mineral at low temperature is a process in which the metal atoms move into the surrounding solution to become aqueous cations, accompanied by a step-wise decrease in the metal to sulphur ratio of the remaining solid phase. When sulphur is finally left over, it undergoes a series of reactions to be oxidized, finally to sulphate ion [8] . This expression may be presented for a simple sulphide $Me\,S$, by the following reactions :

$$Me\,S \longrightarrow Me^{2+} + S + 2e^- \tag{1}$$

(This reaction, being the slowest, is considered as the main reaction which determines the electrode potential).

$$4S + 6OH^- \longrightarrow S_2O_3^{2-} + 2S^{2-} + 3H_2O \tag{2}$$

$$S^{2-} \longrightarrow S + 2e^- \tag{3}$$

$$2S_2O_3^{2-} \longrightarrow S_4O_6^{2-} + 2e^- \tag{4}$$

$$S_4O_6^{2-} \longrightarrow \cdots \longrightarrow SO_4^{2-} \tag{5}$$

Furthermore, metal cations may precipitate at the surface of the mineral if the conditions prevail.

$$Me^{2+} + S_2O_3^{2-} \longrightarrow Me\,S_2O_3$$

$$Me^{2+} + SO_4^{2-} \longrightarrow Me\,S\,O_4$$

$$Me^{2+} + 2OH^- \longrightarrow Me(OH)_2$$

$$Me^{2+} + 3OH^- \longrightarrow Me(OH)_3^-$$

$$Me^{2+} + CO_3^{2-} \longrightarrow Me\,CO_3 \quad , \quad \text{etc.}$$

In addition to the above mentioned reactions, one can intentionally change the electrochemical potentials of the sulphide minerals, in a way similar to the second or third kind electrodes, by the additions of precipitating e.g. Na_2CO_3 , $K_2Cr_2O_7$, $Ca(OH)_2$ or complexing (NaCN) reagents, or by the inhibition of the oxidizing effect of air, (Na_2S, $NaHSO_3$, $FeSO_4$).

It has been established that in the absence of oxygen the fresh surface of a sulphide mineral is wettable to some extent. Oxygen promotes the dehydration of the mineral surface, thus facilitating the penetration of xanthate molecules and their consequent fixation. After long exposure to oxygen the flotation properties af a surface become worse [10, 11] . In addition to the previous description of chemisorption, xanthate coating occurs mainly, in flotation conditions by metathetic reactions between the surface oxidation products and xanthate ions [12, 13] . Hence, the laws governing the equilibria in solutions, between the competing ions should be considered. The significance of quantitative results may however be different for the reactions occurring on the surface. These reactions are characterized by being often irreversible or only slowly reversible and specific for each substrate. For example the solubility of the surface compound, formed by the adsorption of xanthate on pyrite, is completely different from the solubility of the bulk ferrous xanthate; the latter is easily dissolved in water, while the former is insoluble in both water and benzene. Only hydroxyl ions, which interact very actively with iron cations, can replace xanthate anions existing on the surface of pyrite in the form of a surface compound.

Since quantitative data can be obtained only for the bulk reactions, some modifications are required for their applications to surface reactions.

2.2 Hydroxyl Ion as a Depressant and Critical pH Values

The addition of alkali changes the concentration of hydrogen and hydroxyl ions in a solution, and may also change the concentration of heavy metal ions. The influence of the hydroxyl ion on the flotation of sulphide minerals with xanthates and other sulfhydryl collectors showed that, there is a critical pH value below which any given mineral will float and above which it will not float. This critical pH value depends on 1) the mineral, 2) the collector and its concentration, and 3) temperature.

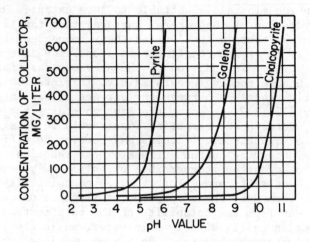

Figure 5. Relationship between concentration of sodium
diethyl dithiophosphate and critical pH value
[14] .

Figure 5 is a typical of the curves of critical pH values determined using the captive bubble technique, in solutions of sodium
diethyl dithiophosphate [14] . Flotation of pyrite, galena and chalcopyrite is possible under the conditions existing to the left of
their respective curves. Table 1 gives values under a variety of
conditions [14, 15] . The quantities of collector used in the tabulated tests were molar equivalents.

Table 1. Critical pH values.

| Mineral | Potassium ethyl xanthate a | | | Sodium Aerofloat A b,f | Sodium diethyl dithio-carbomate c,f | Potassium isoamyl xanthate d,f | Potassium di-n-amyl dithio-carbomate e,f |
	Room temp.	40°F	95°F				
Arsenopyrite	8.4						
Bornite	13.8						
Chalcocite	>14.0						
Chalcopyrite	11.8	13.0	10.8	9.4	>13	>13	>13
Covellite	13.2						
Galena	10.4	10.8	9.7	6.2	>13	12.1	>13
Marcasite	11.0						
Pyrite	10.5	10.2	10.0	3.5	10.5	12.3	12.8
Pyrrhotite	6.0						
Sphalerite	g			g	6.2	5.5	10.4
Sphalerite, R	13.3						
Tetrahedrite	13.8						

a. 25 mg. per li.
b. 32.5 mg. per li
c. 26.7 mg. per li.
d. 31.6 mg. per li.
e. 42.3 mg. per li.
f. Room temperature

g. No contact possible at any pH without activation.
h. The values herein for normal concentrations of collector, as given below. Increase in collector concentration, increases the critical pH.

R = Resurfaced with Cu++.

Barsky suggested that the plot of log concentration of collector vs. pH should give a straight line of unit slope. As shown in Figure 6, the relationships actually obtained are :

$$(X^-) = m \, (OH^-)^n$$

where X^- denotes xanthate ions (or other thiocollectors)and m and n are constants [14,16] .

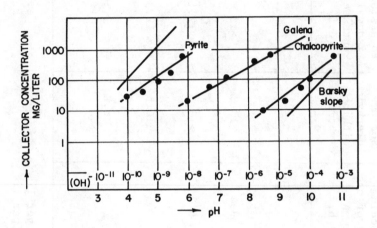

Figure 6. Change in Critical pH values vs. the concentrations of sodium diethyl dithiophosphate, being compared with the theoretical Barsky unit slope [16].

The stability of reagent coatings can be described by their solubility products. In order to compare simply, it should be assumed that the previous surface coating produced by the effect of oxygen of air, is in the hydroxide form.

$$Me \, S + \frac{1}{2} \, O_2 + H_2O \longrightarrow Me \, (OH)_2 + S$$

Then, an equilibrium reaction should occur by the competition of xanthate with hydroxyl ions.

$$Me \, (OH)_2 + 2X^- \longrightarrow Me \, X_2 + 2OH^-$$

The equilibrium constant of this reaction can be written in the following form :

$$\frac{Km_2(OH)_2}{Km_2X_2} = \frac{(OH^-)^2}{(X^-)^2}$$

If the concentration of xanthate is 10^{-4} M, the critical pH may be calculated, for a divalent metal :

$$pH = 10 + \frac{1}{2} \left[pK\,Me\,X_2 - pK\,Me\,(OH)_2 \right] \text{ and for a monovalent metal:}$$

$$pH = 10 + pK\,Me\,X - pK\,Me\,OH$$

This relationship is nevertheless a good approach for guessing the critical pH of simple metal sulphides. The solubility products of some metal hydroxides and ethyl xanthates are given in Table 2. with their respective critical pH's calculated for a xanthate concentration of 10^{-4} M.

Table 2. Solubility products and the calculated critical pH,s

Me S	Me(OH)$_2$; pK	Me X$_2$; pK	Crit. pH
Pb S	Pb(OH)$_2$: 15.2	Pb X$_2$: 16.7	10.7
Cu S	Cu(OH)$_2$: 18.8	Cu X$_2$: 24.2	12.7
Ni S	Ni(OH)$_2$: 17.2	Ni X$_2$: 12.5	7.7
Zn S	Zn(OH)$_2$: 16.5	Zn X$_2$: 8.2	5.9
Fe S	Fe(OH)$_2$: 15.1	Fe X$_2$: 7.1	6.0

Me$_2$ S	Me$_2$O ; pK	Me X ; pK	Crit. pH
Ag$_2$ S	Ag$_2$O : 7.7	Ag X : 18.6	20.9
Cu$_2$ S	Cu$_2$O : 14.8	Cu X : 19.3	14.5

(Data related to metal xanthates are obtained from the references [16] and [17]).

Critical pH values of Table 1, and Table 2 are differing a little and may be accepted as being within the experimental error. Obviously the effect of oxygen, redox reactions induced by the solution or substrate and irreversibility in adsorption etc. are neglected. For example, the critical pH values of marcasite, pyrite and pyrrhotite are different from each other, and can not be explained by a simple metathetic equilibrium.

2.3 Cyanide As A Depressant

Cyanides have been very widely used in the selective flotation of copper-zinc and polymetallic lead-copper-zinc ores, as a depressant for sphalerite, pyrite and certain copper sulphides.

The use of cyanides is made difficult by their high toxicity, relatively high cost and their ability to depress and dissolve gold, thus reducing its extraction into the froth product. In spite of these disadvantages, cyanides are widely used, and attempts to replace them by other depressants have not proved successful in many cases.

In various quantities, cyanides are capable of depressing the flotation of zinc, copper, iron, silver, gold, mercury, cadmium and nickel minerals. The ions of these metals form extremely stable and complex compounds with cyanide. It has been observed that minerals with metallic ions which do not form such compounds with cyanides (lead, bismuth, tin, antimony and arsenic) are not depressed by cyanides [18] .

In order to compare, the stabilities of some metal cyanide complexes their dissociation constants are given in Table 3 [4, 16] .

Table 3. Dissociation constants of metal-cyanide complexes

Metal-Cyanide Complexes	Kd
$Ag(CN)_2^-$	1.8×10^{-19}
$Cu(CN)_2^-$	2×10^{-24}
$Au(CN)_2^-$	5×10^{-39}
$Zn(CN)_4^{2-}$	1.2×10^{-18}
$Cd(CN)_4^{2-}$	1.4×10^{-19}
$Ni(CN)_4^{2-}$	10^{-22}
$Hg(CN)_4^{2-}$	4×10^{-42}
$Fe(CN)_6^{4-}$	10^{-35}
$Fe(CN)_6^{3-}$	10^{-42}

The interrelation of the pH value and the depressant properties of cyanide solutions has been established for collectors of xanthate type. Typical curves are shown in Figure 7. Flotation is possible to the left of the curves. For instance, in a solution containing 25 mg. per liter of potassium ethyl xanthate and 30 mg. of sodium cyanide at a pH value of 7.5, chalcopyrite will float but not pyrite. The flotation of galena is not affected, at pH = 10, one can selectively float galena by using cyanides [14] .

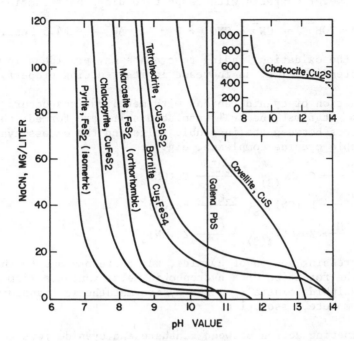

Figure 7. Contact angles for several sulfide minerals.
Bubble contact and flotation possible to left
of the curve (solution contains 25 mg/L potassium ethyl xanthate) [14] .

The effectiveness of Na CN is however decreasing strongly when
pH is lowered; as a result of the hydrolysis of cyanide ion.

$$CN^- + H_2O = HCN + OH^-$$

In order to avoid the formation of HCN, which is a toxic gas,
it is advisable to work at pH higher than the pK_a of hydrocyanic
acid, which is 9.3 . This would reduce the danger to health and
checks unproductive consumption of the reagent.

The reducing property of cyanide is weak. Cyanogen being the
first product of its oxidation :

$$2HCN = (CN)_2 + 2H^+ + 2e^- \quad , \quad E^o = -0.37 \text{ Volt}$$

$$(CN)_2 + 2e^- = CN^- \quad , \quad E^o = -0.18 \text{ Volt}$$

It is however unstable with respect to disproportionation [19].

$$(CN)_2 + 2OH^- = CN^- + CNO^- + H_2O \quad , \quad \Delta G^o = -34.5 \text{ kcal.}$$

Hence, the oxidation leading to cyanate may provide a lower electrode potential, i.e. an increase in its reducing property.

The reaction of cyanide anions with metal cations occurs in two stages : at first, neutral insoluble salts are formed, then a reaction occurs between the insoluble salts and the excess cyanide, forming soluble cyanide complexes, e.g. :

$$Cu^+ \xrightarrow{+CN^-} Cu\,CN_{(S)} \xrightarrow{+CN^-} Cu(CN)_2^-$$

$$Fe^{2+} \xrightarrow{+2CN^-} Fe(CN)_{2(S)} \xrightarrow{+4CN^-} Fe(CN)_6^{4-}$$

$$Zn^{2+} \xrightarrow{+2CN^-} Zn(CN)_{2(S)} \xrightarrow{+2CN^-} Zn(CN)_4^{2-}$$

The depressing action of cyanide, was therefore due to the prevention of the formation of a hydrophobic metal xanthate film at the surface of sulphide minerals either by its metathetic transformation or by its complete dissolution.

The competing action between xanthate and cyanide ions was also influenced by the concentrations of the cations involved in complex formation.

$$Cu\,X + 2CN^- = Cu(CN)_2^- + X^- \quad , \quad K = 2.5 \times 10^4$$

$$\frac{(X^-)}{(CN^-)^2} = \frac{K}{Cu(CN)_2^-}$$

Thus the required cyanide concentration should increase in the same direction as the concentration of soluble copper ions. By the intentional addition of $Cu\,SO_4$ to the pulp, one may nullify the effect of cyanide.

$$Cu^{2+} + 3CN^- \longrightarrow Cu(CN)_2^- + \frac{1}{2}(CN)_2 \quad , \quad K = 10^{29.4}$$

Nevertheless a similar reaction with xanthate ions occurs in the presence of excess cupric ions, leading to a decrease in xanthate ion activity,

$$Cu^{2+} + 2X^- \longrightarrow Cu\,X + \frac{1}{2}X_2 \quad , \quad K = 10^{24.2}$$

but the produced dixanthogen adsorbing onto the allready hydrophobized surfaces or by reacting with the sulphide minerals fortifies the bubble contact.

$$PbS + X_2 \longrightarrow PbX_2 + S$$

2.4 Sulphide As A Depressant

Sodium sulphide additions are used in selective flotation for three distinct purposes [16] .

1. To sulphidize the oxidized metallic sulphides.

2. To improve selectivity in flotation of sulphide minerals by a combined control of pH and sulphide ion concentration.

3. To decompose the xanthate coating on the bulk floated sulphide minerals, ahead of their subsequent selective reflotation.

The first purpose mentioned above, should be considered as an activation process. It is well known that the base metal oxide minerals, such as cerussite, anglesite, malachite, etc., do not float satisfactorily by xanthates. The surface of these ionic minerals are strongly hydrated and the hydrophobic xanthate coating is not sufficiently adherent to withstand the hydrodynamic flotation conditions.

In order to decrease the hydration at these surfaces, sulphidization is required. The result of sulphidization is the passage of sulphur ions into the crystal lattice of the oxidized minerals, replacing sulphate or carbonate ions and transforming the surface of the mineral into a sulphide. The following reaction takes place when sodium sulphide reacts with cerussite [18] :

$$PbCO_3 + S^{2-} = PbS + CO_3^{2-} , \quad K = 6 \times 10^{14}$$

The sulphide film on the oxidized mineral surface is not particularly stable; as the pulp alkalinity increases, colloidal lead sulphide formation becomes more rapid and the film flakes off the surface. The loss of sulphidized coating by attrition is counteracted in flotation practice by making stage additions of $Na_2S \cdot 9H_2O$ instead of a single large addition.

There is an optimum pH value for cerussite (from 9.5 to 10.0), at which there is a maximum adsorption of sulphur ions.

In contrast to cerussite, zinc carbonate (smithsonite) is less susceptible to sulphidization. Experiments show that sulphidization is effective only in a hot pulp (at temperature up to $60^{o}C$).

Eventhough sulphidization is a prerequisite for the flotation of oxidized base metal sulphides, the excess of sulphide ions act as a depressant. The adsorption of sulphide anions on the surface

of sulphide minerals leads to an increase in the negative charge
which prevents adsorption of other anions, including collector
anions; this results in depression of sulphide minerals.

Figure 8, shows the data of Work and Cox for the relationship
between the pH value and the concentration of sodium sulphide required to prevent flotation in the presence of 25 mg. per liter of
the collector potassium ethyl xanthate [14] .

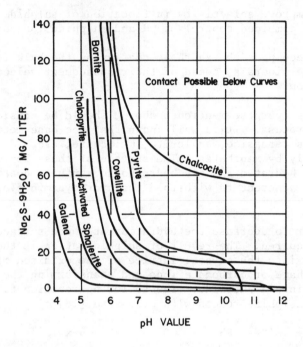

Figure 8. Relationship between pH value and concentration
of sodium sulphide necessary to prevent contact
at surfaces of various sulphide minerals [14] .

The comparison with Figure 7 shows that copper minerals are
depressed by the sulphide in the same order as by cyanide. Pyrite
is relatively less susceptible to sulphide but galena is very sensitive.

Alkali controls the concentration of the sulphide and hydro-
sulphide ions and exercises its own depressant effect above the
critical pH value. The competition between sulphide and xanthate
ions is also dependent of the hydroxyl ion concentration.

$$S^{2-} + H_2O = HS^- + OH^- \quad , \quad K_b = \frac{Kw}{K_2} = 5$$

$$Pb X_2 + HS^- + OH^- = Pb S + 2X^- + H_2O \quad , \quad K = \frac{K_{Pb X_2}}{K_{PbS} \cdot K_b} = 4 \times 10^{10}$$

$$(HS^-) = 2.5 \times 10^{-11} \frac{(X^-)^2}{(OH^-)}$$

It should be recalled that Na_2S being completely hydrolysed, at pH's above 7, (pK$_1$ of H_2S is 7 and pK$_2$ = 14.7); H_2S formation is negligible $Na_2S + H_2O = Na HS + Na OH$.

The concentration of Na_2S may be considered to be equal to (HS^-), which is the dominant sulphur species at these conditions.

In order to prevent the formation of $Pb X_2$ coating on the surface of galena, one may calculate the required concentration of sodium sulphide, at alkaline pH and at a xanthate concentration e.g. pH = 8, $(X^-) = 10^{-4}$ M : $(HS^-) \geqq 2.5 \times 10^{-13}$ M.

The result obtained by the above calculation is obviously very low and could not be compared with the results (at aerated conditions) of Wark and Cox. Even fresh galena would provide such a low value.

$$Pb S + H_2O = Pb^{2+} + HS^- + OH^- \quad , \quad K = K_{sp} \cdot K_b = 5 \times 10^{-28}$$

e.g. at pH 8, $(HS^-) = 2 \times 10^{-11}$ M. This result, strongly supports the necessity of the oxidation of sulphide ions by air prior to flotation. The depressing action of sodium sulphide is particularly strong in the flotation of unoxidized sulphide minerals by xanthate. Even a fraction of a milligramme of Na_2S per liter exhibits depressing action on the flotation of unoxidized galena. This behaviour is analogous to the flotation of sulphide minerals without any interaction with oxygen. Sulphide minerals depressed by sodium sulphide regain flotability if sufficient time is allowed for the dissolved oxygen to react with the sulphide [5] . This does not only reduce the depressing activity of sulphide ion but provides also some natural hydrophobicity.

The third mode of utilizing sulphide additions is to destroy the hydrophobic collector coating on the bulk-floated minerals. As the complex sulphide ores available for treatment become more finely disseminated, bulk flotation of sulfides becomes the preferable treatment. Any subsequent retreatment of the bulk concentrate necessitates a partial or a complete destruction of the initially induced hydrophobic coating. The decomposition technique is employed in industrial practice for the separation of Cu-Mo and Cu-Ni minerals; as outlined in the flowsheed in Figure 9 [16] .

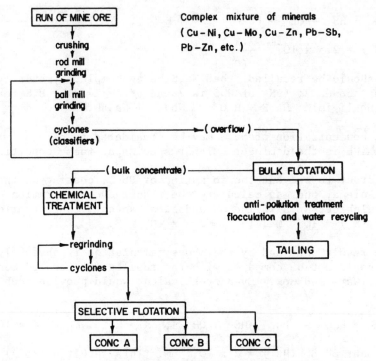

Figure 9. Bulk flotation of sulfides followed by chemical treatment regrinding, and finally, selective separations by flotation. Adopted for
finely disseminated complex sulfide ores, for
example; Cu-Mo, Cu-Ni, Ni-Co-Cu, etc.

Beside being a potential determining ion and thus lowering the
electrode potentials of metal sulphides by its precipitating action,
sulphide ions in appreciable quantities decrease the redox potential
of the media and prevent the oxidation of cyanide to cyanogen and
xanthate to dixanthogen, i.e. oxygen of air will oxidize sulphide
ions before the others.

The redox potentials calculated from the free energies of
H_2S, HS^- and $S^=$ do not reflect the actual influences, because of the
dissolution of free sulphur in excess sulphide to form polysulphides,
and in alkali to form thiosulphate and sulphide. Nevertheless, the
following equations and the related potentials may give some idea
of the reducing power of sodium sulphide additions [19] .

$$\text{pH} < 7 \; : \quad S + 2H^+ + 2e^- = H_2S \qquad , \qquad E^o = -0.14 \text{ Volt}$$

$$\text{pH} > 7 \; : \quad S + H^+ + 2e^- = HS^- \qquad , \qquad E^o = -0.06 \text{ Volt}$$

$$S + 2e^- = S^{2-} \qquad , \qquad E^o = -0.48 \text{ Volt}$$

$$\text{in} \quad NaOH = 1 \text{ Molal} \; S_2^{2-} + 2e^- = 2S^{2-} \quad , \quad E_B^o = -0.48 \text{ Volt}$$

$$4S + 4OH^- = 2HS^- + S_2O_3^{2-} + H_2O \qquad , \qquad \Delta G^o = -27.5 \text{ kcal.}$$

3. ELECTROCHEMICAL STUDIES

3.1 Galena-Oxygen-Xanthate Flotation System [20, 21]

It has been proved in a long series of papers by Plaksin and his co-workers that oxygen is an essential participant in the flotation of galena by xanthates [5, 10] . Furthermore, it has been claimed that (a) the adsorption step requires oxygen, (b) oxygen affects the wettability of galena itself, and (c) when xanthate is taken up by galena the reaction occurs at localized patches and on some faces or some grains more than on others also that the heavier the deposition, the more pronounced the non-uniformity.

In addition, an appreciable time is required for interaction of galena with oxygen, either before or during its contact with xanthate solution, even when the amount of oxygen available is for greater than that required to form a monolayer on the galena [22] . This observation has a bearing on the claims of Nixon [23] . Golikov [24] and others that galena catalyses the oxidation of xanthates to dixanthogens which then improves flotation.

The use of electrode potentials for elucidating reactions is, of course, well known. For example the effect of xanthate on the potential of galena electrodes was studied by Lintern and Adam [25] a long time ago in 1935. Polarization curves are often informative in systems where it is not possible to establish a well defined thermo-dynamically reversible electrode potential - for example, in the system of $Pb/PbS/H_2O$. Such electrodes are often not at equilibrium, but in a state exhibiting a so-called "mixed-potential" which depends on the relative rates of simultaneous oxidation and reductions at the electrode.

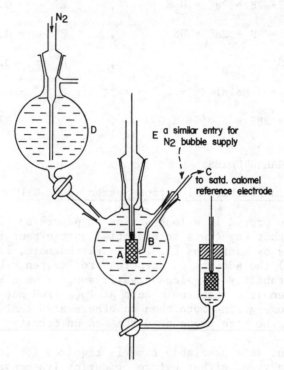

Fig. 10 Diagram of the apparatus.

Tolun and Kitchener [20] constructed a special all - glass
cell (Figure 10) which permitted measurement of electrode potentials
at a working electrode, A, by means of a Luggin capillary, B, and
a saturated calomel reference electrode, C, in the absence of air.
Solutions could be deaerated in a side vessel, D, by passage of
"oxygen-free" nitrogen. A stream of this gas could also be passed
through the elctrode vessel, and its bubble could be applied to the
electrode when desired by means of a rotatable bent capillary jet,E.
With this apparatus, the investigators were able to measure the quantities of electricity required to surmount the oxidation-reduction
steps in the polarisation curves. Determinations were carried out
with a base solution of borax buffer at pH 9.

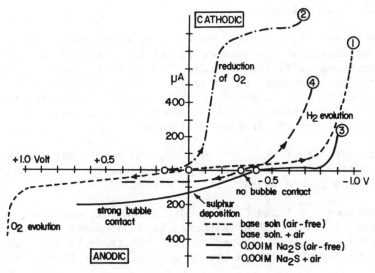

Figure 11. Polarograms for Na$_2$S solutions on platinum

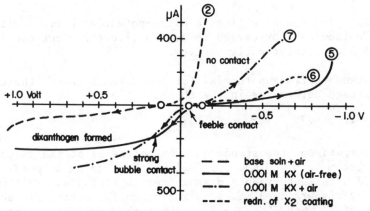

Figure 12. Polarograms for potassium ethyl xanthate on
platinum.

i. Electrolytic oxidation of sulphide and xanthate ions on platinum.

The polarograms collected in Figure 11. lead to the following
conclusions :

(a) The first oxidation product of sulphide ions is elementary
sulphur, which makes the (otherwise hydrophillic) platinum
strongly hydrophobic (curve 3). (This result is concordant
with the hydrophobicity created by the slight oxidation of

metal sulphides).

(b) With Na_2S plus air (curve 4), a mixed-potential curve is obtained, but the current is not the algebraic sum of the separate oxidation and reduction currents (curve 2 and 3); the oxygen-reduction process is strongly poisened by the sulphur. (The cathodic part of curve 2, indicates the first reduction step of oxygen to hydrogen peroxide). The polarograms in Figure 12 shows :

(c) In absence of air, 10^{-3} M potassium ethyl xanthate (KX) lowers the potential of a platinum electrode to - 0.12 V, but no bubble contact can be formed.

(d) Between - 0.1 and 0.0 V onodic oxidation commences, with slow formation of dixanthogen (X_2) and onset of feeble bubble contact. Above 0.0V the electrode becomes coated with dixanthogen and strong bubble contact is observed.

(e) The dixanthogen is cathodically reduced back to xanthate (curve 6), contact being retained as long as it is present.

(f) Air plus xanthate gives a mixed-potential curve 7, with "poisoning" for oxygen reduction (i.e the electrode is no longer catalytically active).

The standard redox potential for the dixanthogen-xanthate system has been determined in the absence of air.

$$X_2 + 2e^- = 2X^- \quad , \quad E^o = - 0.081 \text{ V}$$

For comparison, the standard redox potential of the sulphur-sulphide system is - 480 mV. It follows that sodium sulphide solution should be able to reduce dixanthogen to xanthate. This was confirmed by dispersing 3 mg. of dixanthogen (by ultrasonic action) in 100 ml. of 10^{-3} M, Na S (pH 9.2) : the turbid solution on standing became completely clear after 1/2 hour, and ultra-violet spectrophotometry showed the quantitative formation of xanthate (with its maximum at 3000 A^o). The sulphur formed by oxidation remained in solution as disulphide (S_2^{2-}). The equilibrium constant calculated for the related equation clearly supports this reaction.

$$X_2 + 2S^{2-} \longrightarrow S_2^{2-} + 2X^- \quad , \quad K = 10^{13.5}$$

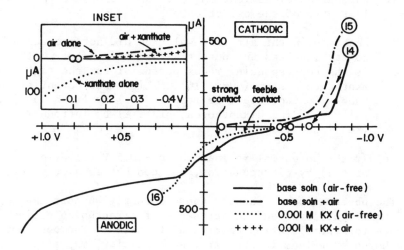

Figure 13. Polarograms at a galena electrode.

ii. Reactions at a galena electrode

The polarograms in Figure 13 show :

(a) In the presence of air, the potential shifted strongly to
less negative values(- 0.1 V, curve 15), indicating chemisorption of oxygen. On cathodic polarization, the galena
electrode did not readily reduce dissolved oxygen.

(b) Anodic polarization produced sluggish oxidation (curve 14)
presumably first to a basic thiosulphate [26] , which may
be represented by the following overall equation,

$$2Pb\,S + 5H_2O = Pb(OH)_2 \cdot Pb\,S_2O_3 + 8H^+ + 8e^- \, , \quad E^o = 0.505 \text{ V}$$

resulting from the following reactions :

| galena n-type with excess Pb^o | $\longrightarrow$ | galena p-type with excess S^o | $+ Pb^{2+} + 2e^-$ |

$$4S^o + 6OH^- \longrightarrow 2S^{2-} + S_2O_3^{2-} + 3H_2O$$

$$Pb^{2+} + S^{2-} \longrightarrow Pb\,S$$

$$2Pb^{2+} + 2OH^- + S_2O_3^{2-} \longrightarrow Pb(OH)_2 \cdot Pb\,S_2O_3$$

(c) Further anodic oxidation, obviously produced $Pb\,SO_4$ as the last step of the oxidation.

(d) Xanthate in the air free solution did not produce a bubble contact, unless the potential was raised to - 0.13 V, which is the corresponding redox potential for the xanthate-di-xanthogen system (curve 16). This mixed potential and bubble contact could be restored (i) by fresh anodization, (ii) by adding dixanthogen, or (iii) by introducing air (see inset).

(e) If the potential is lowered to - 0.5 V (i) by cathodization or (ii) by nitrogenation, the contact is slowly destroyed.

The observations presented above strongly support the importance of mixed potential and the presence of dixanthogen beside lead xanthate for the flotation condition of galena. The consumption of dixanthogen by galena in the air free condition, may simply be described by the following reaction :

$$X_2 + Pb\,S \longrightarrow Pb\,X_2 + S$$

The lead xanthate film alone produced in this way, does not provide a floatable condition.

3.2 Favourable Area for Flotation of Galena on a Potential-pH Diagram

i. Galena-oxygen-water system :

The minimum potentials required for the oxidation of galena can be calculated from the following equations :

$$2Pb\,S + 5H_2O = Pb\,S_2\,O_3 + Pb(OH)_2 + 8H^+ + 8e^-$$

$$Eh = 0.505 - 0.059\ pH$$

$$2Pb\,S + 7H_2O = 2Pb(OH)_2 + S_2O_3^- + 10\,H^+ + 8e^-$$

$$Eh = 0.632 - 0.0737\ pH + 0.007\ log\ (S_2O_3^{2-})$$

$$2Pb\,S + 7H_2O = 2H\,Pb\,O_2^- + S_2O_3^{2-} + 12H^+ + 8e^-$$

$$Eh = 0.841 - 0.0885\ pH + 0.015\ log\ (H\,Pb\,O_2^-) + 0.007\ log\ (S_2O_3^{2-})$$

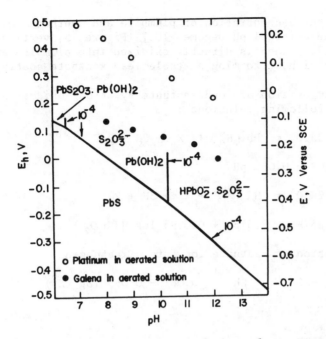

Figure 14. Potential-pH diagram of the galena-oxygen-
water system [21] .

The equilibrium lines collected in Figure 14 lead to the following conclusions :

(a) Galena is an unstable mineral in oxygen-containing water.

(b) The mixed potential of the galena electrode (closed circles)
is determined by the equal rates of the cathodic reduction
of oxygen and anodic oxidation of galena.

ii. Galena-oxygen-xanthate system :

The electrochemical reaction leading to the production of lead
xanthate is :

$$2Pb\,S + 3H_2O + 4X^- = 2Pb\,X_2 + S_2O_3^{2-} + 6H^+ + 8e^-$$

$$Eh = 0.194 - 0.044\ pH + 0.007\ \log\ (S_2O_3^{2-}) - 0.0295\ \log\ (X^-)$$

The stability of lead xanthate is determined by the following
equations [27] :

$$Pb\,X_2 + 3OH^- = H\,Pb\,O_2^- + 2X^- + H_2O \quad , \quad K = 10^{-2}$$

Keeping the concentration of plumbite and xanthate ions equal to 10^{-4} M, the critical pH becomes 10.7 (Figure 15, vertical line). Beyond pH 10.7, galena is directly oxidized into plumbite and thiosulphate ions without forming a stable lead xanthate coating.

The decomposition of lead xanthate by strong oxidation is determined by the following relations :

$$Pb\,X_2 + 2H_2O = Pb(OH)_2 + X_2 + 2H^+ + 2e^-$$

$$Eh = 0.800 - 0.059\ pH$$

$$Pb\,X_2 + 2H_2O = H\,Pb\,O_2^- + X_2 + 3H^+ + 2e^-$$

$$Eh = 1.225 - 0.0885\ pH + 0.0295\ \log H\,Pb\,O_2^-\ )$$

and the oxidation of xanthate ion into dixanthogen

$$2X^- = X_2 + 2e^- \quad , \quad Eh = -0.081 - 0.059\ \log (X^-)$$

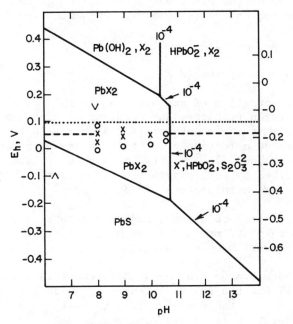

Figure 15. Galena-xanthate-oxygen system.

The equilibrium lines, together with the experimental conditions of strong and feeble bubble contact, are shown in Figure 15.

The bubble adhesion was only obtainable in the area where lead xanthate was found to be stable, and the oxidation of xanthate ion was appreciable.

It is also interesting to observe that the potentials corresponding to strong bubble contact fall in an area parallel to the equilibrium potential of the xanthate-dixanthogen system, and nearly coincide with the equilibrium potential of the dixanthogen-galena system (broken line) :

$$Pb\,S + X_2 = Pb\,X_2 + S \quad , \quad K = 10^{1.45}$$

The first step of the electrode potential of galena in the presence of xanthate ion coincides also with this flotation condition.

$$Pb\,S + 2X^- = Pb\,X_2 + S + 2e^-$$

$$Eh = -0.124 - 0.059 \log (X^-)$$

When the concentration of xanthate ion is 10^{-3} M, Eh becomes 0.053 V. The electron transfer, and chemisorption of xanthate becomes appreciable at these potentials. At higher potentials, the formation of hydrated oxidation products competes with the hydrophobic lead xanthate and finally prevents bubble contact. Such conditions were obtained by full oxygenation or better by the depressing action of potassium chromate at pH 8.

As with oxygen-free solutions, reducing agents, such as sodium sulphite-ferrous sulphate mixture at pH 6.5, producing sufficiently low potentials, prevent bubble contact.

It should therefore be concluded that, there is a limiting range of potential for the system galena-oxygen-xanthate solution at which bubble contact takes place, just as there is a limiting pH range where flotation is possible.

4. SILVER SULPHIDE ELECTRODE AND ITS USE IN CONTROLLING FLOTATION OF AN OXIDIZED LEAD-ZINC ORE

In the sulphidization of oxidized ores for flotation the sulphidization stage is critical, because either too much or too little sodium sulphide gives poorer metallurgy than does an optimum addition [28,29]. The use of silver sulphide as a sulphide ion-selective electrode is the most convenient technique for controlling the operating conditions.

4.1 Silver Sulphide-Xanthate-Oxygen System [30].

A solid microcrystalline silver sulphide of a thickness of about 2 mm. and of a diameter of 10 mm. has been prepared at the laboratory, by melting precipitated silver sulphide with some sulphur in a loosely covered porcelain crucible. During melting, excess sulphur is evaporated and during cooling silver sulphide is solidified as a microcrystalline disc at the bottom of the crucible. Polished surface of this material, examined by reflected light microscopy, should be free of metallic silver. (The non stoichiometry of silver sulphide is well known. It may accommodate appreciable amount (< 0.1 %) of silver or sulphur as solid solution; nevertheless it remains always as an n-type semiconductor; but the excess sulphur containing species have lower conductivities). The solid silver sulphide was cemented to the end of a glass tube with Araldite resin, electrical contact being made to a silver wire by a drop of mercury in the tube [33] .

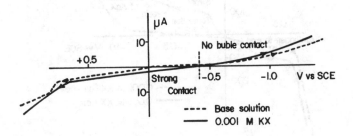

Figure 16. Polarograms at a silver sulphide electrode.

Silver sulphide electrode is known for its sensitivity toward silver or sulphide ions. As these potential determining ions could not be provided appreciably from the surface of silver sulphide ($K_{sp} = 10^{-51}$), the static potential was affected very slowly by oxygen reduction of air. As it is shown in Figure 16, the polarograms in air free or aerated solutions are the same. The static potential established very slowly and reached about − 0.04 V at the end of several hours.

When potassium ethyl xanthate was added to the base solutions, either air free or aerated, the static potential of the electrode is lowered and rested at about − 0.25 V. No dixanthogen formation was observed during the anodic polarization of the electrode and no oxygen reduction during its cathodic polarization. This was an indication that the electrode was free of catalytic activity and should not suffer of the mixed potential effect of air. This effect is clearly shown in Figure 17, when silver metal was used as electrode instead of Ag_2S.

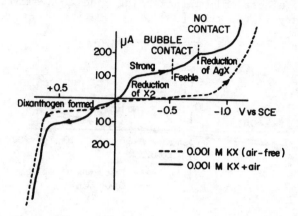

Figure 17. Polarograms for potassium ethyl xanthate on silver.

The calibration curve presented in Figure 18 indicates a satisfactory sensitivity of this electrode towards xanthate ion concentrations down to 10^{-5} M. The potentials obtained were higher in magnitude than the potentials of aerated suspensions of galena or sphalerite.

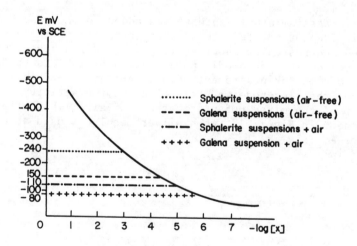

Figure 18. Electrode potentials versus to the concentrations of potassium ethyl xanthate on silver sulphide.

Shortly to say that the electrode should be able to indicate the activities of xanthate ion, when the competing activity of sulphide ion is not appreciable. The following equations support also qualitatively this behaviour :

$$Ag^+ + e^- = Ag \qquad ; \qquad E^O = 0.799 \text{ V}$$

$$Ag X = Ag^+ + X^- \qquad ; \qquad K_{sp} = 4.2 \times 10^{-19}$$

$$E_{SCE} = -0.53 - 0.059 \log (X^-)$$

$$Ag_2S + H_2O = 2Ag^+ + HS^- + OH^- \qquad ; \qquad K_{sp} \cdot K_w \cdot K_2^{-1} = 10^{-50}$$

$$Ag_2S + H_2O + 2e^- = 2Ag + HS^- + OH^- \qquad ; \qquad E_{SCE}^O = -0.9 \text{ V}$$

$$E_{SCE} = -0.5 - 0.03 \text{ pH} - 0.03 \log (HS^-)$$

$$(X^-)^2 \gtrsim 10^{13} \text{ } OH^- \text{ } HS^-$$

(i.e. in order to measure 10^{-5} M xanthate ion activity at pH 9, HS^- should be smaller than 10^{-18} M.

4.2 Concentration of an Oxidized Lead-Zinc Ore

The investigated sample was received from a mine situated near the town of Kayseri (Zamanti district). The main minerals identified in this ore sample :

Smithsonite	Calcite	Clay minerals
Hydrozincite	Sericite	Limonite
Cerussite		

The main operations used for its concentration :

i. Crushing and grinding to - 48 mesh.

ii. Desliming of fines (- 10 μ m) by hydrocyclone.

iii.Conditioning and flotation of cerussite.

iv. Desliming by decantation.

v. Flotation of smithsonite.

The results of two successful tests are presented in the following Tables :

Table 4. Concentration results of a rich ore [31] .

Product	Weight g.	Grade, %		Recovery, %	
		Pb	Zn	Pb	Zn
Lead concentrate	84	64.3	1.88	89.3	1.4
Zinc concentrate	191	0.83	43.6	2.6	72.9
Tailing	222	0.92	12.6	1.9	13.4
Slimes	97	3.88	14.5	6.2	12.3
Feed	494	12.2	23.1	100.0	100.0

Table 5. Concentration results on the coarse tailing of the gravity concentration by jigging [32] .

Product	Weight g.	Grade, %		Recovery, %	
		Pb	Zn	Pb	Zn
Lead concentrate	126	58.0	2.0	84.0	2.1
Zinc concentrate	200	1.0	50.5	2.3	84.1
Tailing	514	1.8	2.7	10.4	11.7
Slimes	160	1.8	1.5	3.3	2.0
Feed	1000	8.7	12.0	100.0	100.0

i. Flotation of cerussite

The selective flotation of the oxidized lead minerals is carried out by treating them first with sodium sulphide and then by using a xanthate collector, the operation is accomplished in a similar way to galena flotation. The conditions of a successful flotation test is summarized in the following Table :

Table 6. Flotation conditions of cerussite

			E_{SCE} ; mV
Pulp : solid / liquid	= 1/5 ; pH = 9		
Addition of $Na_2S \cdot 9H_2O$	:	200 mg/1	$-$ 470
I. conditioning	:	5 minutes	$-$ 140
Addition of K A X	:	150 mg/1	$-$ 250
II. conditioning	:	4 minutes	$-$ 140
Flotation	:	10 minutes	$+$ 20

The increase of the potential during the first conditioning indicated the concumption of sulphide ions by lead corbonate mine-rals. At the end of 5 minutes, the potential was just a little higher than that of a galena suspension in the air free solution (Figure 18), indicating the end point of the metathetic reaction.

$$Pb\,CO_3 + HS^- = Pb\,S + HCO_3^- \quad ; \quad K = 3 \times 10^{10}$$

The addition of K A X lowered the potential to an expected value and at the end of the conditioning, the concentration of xanthate ion falls to about 10^{-5} M. During the flotation, lead minerals and xanthate ions are removed with the froth, and the po-tential increased as a result of the oxidizing action of air in the pulp.

ii. Flotation of smithsonite

The flotation of oxidized zinc minerals is carried out selec-tively by treating them with sodium sulphide and a fatty amine. The conditions of a successful test is summarized in the following Table :

Table 7. Flotation conditions of smithsonite

			E_{SCE} ; mV
Addition of $Na_2S \cdot 9H_2O$	:	2000 mg/1	$-$ 680
pH = 12 (adjusted with Na OH)			
Addition of tallowamine acetate	:	100 mg/1	$-$ 680
(together with pine oil and	:	15 mg/1	
kerosene in emulsified form)	:	25 mg/1	
Conditioning	:	4 minutes	$-$ 680
Flotation	:	10 minutes	$-$ 650

The potential, indicating the activity of sulphide ion, did not change during the conditioning. Hence, no appreciable consumption of sulphide by zinc minerals was observed. Nevertheless, the data on the solubility products of zinc sulphide $(10^{-21.5})$ and zinc hydroxide $(10^{-16.5})$ support the formation of zinc sulphide at the surface of zinc minerals. Thus the dissolution as zincate ion was prevented and the particles were selectively floated with fatty amine collectors as if they were sphalerite [4].

$$Zn(OH)_2 + HS^- = ZnS + OH^- + H_2O \quad ; \quad K = 10^4$$

$$Zn(OH)_3^- + HS^- = ZnS + 2OH^- + H_2O \quad ; \quad K = 5 \times 10^6$$

In our conditions (pH = 12), the added sodium sulphide was more then sufficient for the required activity of HS^- (10^{-6} M).

The addition of tallowamine did not alter the potential. The expected adsorption of this collector onto the surface of zinc minerals could therefore not be observable on the potential of the electrode, because of the extreme insolubility of silver sulphide compared to its amine complexes.

In the light of the presented data, it should be concluded that, beside the pH, the continuous control of sulphide and xanthate ions by the silver sulphide electrode is very practical to achieve reproducible results during the flotation of oxidized lead-zinc ore.

ACKNOWLEDGEMENTS

The author thanks the NATO Scientific Affairs Division for award on the collaborative research project, No. 258.81, which enabled him to profit of the valuable advises of Dr. J.A. Kitchener and Dr. G.H. Kelsall (i.e. Dept. Min. Res. Engng.) , and Prof. C.Ek (Universié de Liège). Thanks are also due to Prof.Dr.Z.Doğan (METU) who kindly revised and corrected the manuscript.

REFERENCES

1. Eigeles, M.A. Theoretical Basis of Flotation of Non-Sulphide Minerals (Metallurgizdat, 1950).

2. Dennen, W.H Principle of Minerology (Ronald Press, New York, 1959).

3. Gaudin, A.M. Flotation (Mc Graw-Hill, New York, 1957).

4. Klassen,V.I. and V.A. Mokrousov. An Introduction to the Theory of Flotation (Butterworths, London, 1963).

5. Pourbaix, M. Atlas of Electrochemical Equilibria in Aqueous Solutions. 2'nd Edition (National Association of Corrosion Engineers, Houston, U.S.A., 1974).

6. Garrels, R.M. and C.L. Christ. Solutions, Minerals and Equilibria (Harper and Row Publishers, New York, 1965).

7. Delahay, P.A. A Polarographic Method for the Indirect Determination of Polarization Curves for Oxygen Reduction on Various Metals, Journal of Electrochem. Society 97 No.6 (1950)198-212.

8. Sato, M. Oxidation of Sulfide Ore Bodies,Economic Geology 55 (1960) 928-961 and 1202-1231.

9. Venkatachalam, S. and R. Mallikarjunan. Electrode Potentials and Anodic Polarization of Sulphides, Trans. IMM 79 (1970) C 181-188.

10. Plaksin, I.N. Interaction of Minerals with Gases and Reagents in flotation, Mining Engineering (1959) March 311-324.

11. Fuerstenau,D.W. Froth Flotation, 50'th Anniversary Volume. Chapter 6 by Rogers J. (A.I.M.E. New York, 1962).

12. Mellgren, O. Heat of Adsorption and Surface Reactions of Potassium Ethyl Xanthate on Galena, Trans. A.I.M.E. 235 (1966) 46-60

13. Mellgren, O. and S.L. Lwakatare. Desorption of Xanthate Ions from Galena with Sodium Sulphide, Trans. IMM 77 (1968) C 101-104.

14. Wark, I.W. and A.B. Cox Trans. A.I.M.E. 112 (1934) 189, 245 and 267.

15. Taggart, A.F. Handbook of Mineral Dressing (John Wiley, 1956).

16. Leja, J. Surface Chemistry of Froth Flotation (Plenum Press, New York and London, 1982).

17. Due Rietz, C. Xanthate Analysis by means of Potentiometric Titration, Svensk Kem. Tidskr. 69 (1957) 310-327

18. Glembotskii, V.A., V.I. Klassen and I.N. Plaksin. Flotation (Primary Sources, New York, 1972).

74

19. Latimer, W.M. Oxidation Potentials (Prentice Hall, Englewood Cliffs, 1956).

20. Tolun, R. and J.A. Kitchener. Electrochemical Study of the Galena-Xanthate-Oxygen System, Trans. IMM 73 (1964) 313-322.

21. Toperi, D. and R. Tolun. Electrochemical Study and Thermo-dynamic Equilibria of the Galena-Oxygen-Xanthate Flotation system, Trans. IMM 78 (1969) C 191-197.

22. Beebe, R.R. and C.E. Westley. Contact Angle on Galena as a Function of Oxygen Concentration, Trans. A.I.M.E. 220 (1961) 9-10.

23. Cook, M.A. and M.E. Wadsworth. Hydrolitic and Ion Pair Adsorp-tion Processes on Flotation, Ion Exchange and Corrosion, Proc. 2'nd Int. Congress Surface Activity(Butterworth,London, 1957), III, 228-242, Discussion by Nixon J.C. 369-370.

24. Golikov, A.A. Interaction of Xanthogenate Type Collectors on Sulphide Mineral Surfaces, Soviet J. Non-Ferrous Metals 2 (1961) 715-722.

25. Lintern, P.A. and N.K. Adam. The Influence of Adsorbed Films on the Potential Difference between Solids and Aqueous Solutions with Special Reference to the Effect of Xanthates on Galena, Trans. Faraday Soc. 31 (1935) 564-574.

26. Eadington, P. and A.P. Prosser. Surface Oxidation Products of Lead Sulphide, Trans. IMM 75 (1966) C 125.

27. Fleming, M.G. Effects of Alkalinity on the Flotation of Lead Minerals, Min. Engng. 4 (1952) 1231-1236.

28. Jones, M.H. and J.T. Woodcock. Evaluation of Ion-Selective Electrode for Control of Sodium Sulphide Additions during Laboratory Flotation of Oxidized Ore, Trans. IMM 87 (1978) C 99-105.

29. Jones, M.H. and J.T. Woodcock. Control of Laboratory Sulphidi-zation with a Sulphide-Ion-Selective Electrode, before Flotation of Oxidized Lead-Zinc-Silver Dump Material, Int. J. Mineral Proc. 6 (1979) 17-30.

30. Tolun, R. Silver Sulphide Electrode and Its Use for Controlling the Flotation of an Oxidized Lead-Zinc Ore, Bull. Tech. Univ. İstanbul 36 (1983) 47-55.

31. Imre, M. Development of a Process to Treat an Oxidized Lead-Zinc Ore from Turkey, M. Sc. Thesis, Imperial College of Science and Technology, Dept. Mineral Tech., London (1969).

32. Göksu, M. The Concentration of an Oxidized Lead-Zinc Ore by Flotation and Its Control by Using Silver Sulphide Electrode, M.Sc. Thesis Middle East Technical University, Ankara (1972).

33. Ilter, B. Electrochemical Studies on Sulphide Ion Sensing Electrode, M.Sc. Thesis, Middle East Technical Univ. (1972).

COAL FLOTATION

Gülhan Özbayoğlu

Associate Professor in Mining Engineering
Department of Middle East Technical University,
Ankara-TURKEY

ABSTRACT

Coal flotation of fine material has both economical and environmental benefits by recovering as much coal as possible from
run-of-mine coal and by reducing air and water pollution due to
the elimination of sulfur from clean coal and reduction in the
solid amount which goes into the stream or in recirculating plant
water.

Coal flotation is characterized by the chemical and petrographical composition of coal and its rank. In addition, size and
sp . gr. of coal particles, pulp density, rate and uniformity of
feed, conditioning, type of reagents, pH, presence of clay and
type of flotation machines are the other variables which influence the coal flotation.

This paper reviews most of the parameters affecting coal flotation.

1. INTRODUCTION

Flotation is a physico-chemical method of concentration where-
by minerals are separated from each other or from their associated
gangue. Although flotation was originally developed in the mineral
industry and used commercially as early as 1902, the process has
been gradually extended to other fields, but the coal practice was
not started until 1920 (1). European countries, notably Holland,
has led the way in the application. It has grown more slowly

in the United States than in Europe. In the U.S.A., the production of
flotation concentrate achieved was over 14 million tons which was
about 5 percent of the cleaned coal in 1978 (2,3). The percentage
of clean coal produced by flotation in Europe (England, West Germany,
Poland, U.S.S.R., etc) is substantially higher (4). For example,
in the Soviet Union the amount of coal fines treated by flotation
exceeds 44 million tons in 1979. The share of froth flotation in
the total tonnage of cleaned coal is over 12 percent (5). In Turkey,
about 14-18 percent of the run-of-mine bituminous coal which comes
to the washery is treated by flotation (6).

Benefits, both economic and enviromental, gained by beneficiation has been a necessity in most coal operations requiring some
degree of preparation which is due to the following considerations:

i) Mining of coal by highly mechanized methods has caused an increase
in the fine fraction.
ii) Black water problem which has to be controlled from an environmental point of view.
iii) The need for the production of coal with low sulfur and ash
content leads to an increasing emphasis on flotation as a means of
achieving these ends.
iv) With substantial increase in the value of coal and its cost of
production, the fines previously wasted become a very valuable product.

The fine fraction of a coal stream is that fraction which contains
particles essentially in the -28 mesh size range.This fraction will
normally first appear, in a coal preparation process, as a result
of classification to segregate raw coal into various size components
for beneficiation. More specifically, it could appear in desliming
screen overflow, classifying cyclone overflow, and in other process
circuit components. Therefore, the essential difference between ore
and coal flotation is that for ores the entire tonnage is ground
to flotation size, for coal, however, comminution does not occur,
and only that fine fraction not processed by gravity concentration
is treated by flotation.

Although coal flotation has a tremendous potential, it has not
been widely practiced.
The explanation is given as follows:

i) High cost of flotation process.

Capital and operating costs associated with coal beneficiation are
principally a function of the surface area of the coal being treated.
Since a ton of -28 mesh fine coal has a much greater surface area
than a ton of +28 mesh coarse coal, its beneficiation costs are much
greater. In general, it costs three times as much to wash fine coal

as it does to wash coarse coal (7). Nevertheless, in most cases ,
coal prices are now high enough to justify fine coal beneficiation.
ii) Failure to clean the fines containing high percentage of clay.
iii) Tendency for oversize to be lost in tailings.
iv) Marketing problems of fines.
v) High cost of dewatering.
More machinery and energy is needed to remove moisture from fine
particles due to their large surface area than to remove moisture
from coarse particles.
vi) Difficulty of application to certain coals.
The beneficiation processes based on specific gravity are only
slightly affected by rank of the coal, degree of Oxidation of the
particles, or the chemical quality of the clarified water, the
floatability of the coal is extremely sensitive to all of these
conditions. Most commercial reagents do not react with oxidized coal.
Therefore, most of the oxidized coal processed in flotation cells
reports to the tailings. Low rank coals and higher rank coals that
have become oxidized are not amenable to froth flotation or will
give very low yields (3).

Due to the complexity and heterogeneity of the coals, it is appropriate to survey some of the properties of coal and operational factors
which are of importance in the flotation.

2. STRUCTURE AND COMPOSITION OF COAL AFFECTING COAL FLOTATION

2.1. Chemical and petrographical composition of coal

    Coal is a combustible sedimentary rock formed from plant remains
under the influence of high temperature and pressure. It is characterized by its rank and type. The nature of the original plant material determines the type of coal, while the degree of metamorphism
determines the rank which is the stage reached by a coal from peat
to anthracite in the course of its coalification.

    Coals with their constituents vary in chemical composition from
one to another. Due to their extra ordinarily complex carbon chemistry, these combustible solid minerals can not be represented by
uniquely defined chemical structures. But it is accepted that coals
are comprised of C,H,O,S and N. Structurally, they are composed of
condensed aromatic rings of rather high molecular weight. About 70%
of all carbon atoms are in aromatic rings, but only about 20-25% of
hydrogen atoms are attached to aromatic carbon atoms (8,9). Oxygen,
sulfur and nitrogen are combined in chemically functional groups
such as OH, CO, COOH, NH, CN , S, SH etc., which occur as integral
parts of the original molecules (9,10).

    Due to the high temperature during coalification, some chemical
transformations take place increasing the carbon content of coal

while diminishing of its oxygen, hydrogen, and volatile matter contents. Table 1 shows classification of coals and their properties
as a function of rank (11). As it is seen,rank increases from left
to right. Heating value, sometimes expressed as "Rank Btu" which
tends to increase with rank while moisture content tends to decrease.

One of the most important features of a coal is its internal
capillary structure. The presence of pores and cracks determines the
specific surface area reacting with flotation reagents. During chemical reaction or adsorption processes, reagents have to pass through
these very small capillaries, and specific adsorption processes are
followed by condensation of reagents in micropores. This capillary
condensation takes place as a result of Van der Waals forces.There-
fore, the adsorptive properties of coal depend on the nature of the
pore structure (12,13).

It is also known that the equilibrium moisture contents of coals
are dependent on their porous structure.Coal becomes more   porous
when its rank decreases and becomes more difficult to treat by froth-
flotation. Brown coal and lignite have a strongly hydrophilic character and a high intrinsic moisture content. In the transition from
brown coal and lignite to bituminous coal, the inherent moisture
content decreases from 45 to 2-3 percent and the coal substance
becomes more hydrophobic (14).

Four distinct "ingredients" or "rock types" (lithotypes) were
distinguished with the unaided eye in banded bituminous coals.Four
terms, vitrain, clarain, durain, and fusain were first described
by Stopes (15). The petrographic units of lithotypes which do not
appear to be capable of further resolution under the microscope were
called "macerals". Each maceral has been given a special name such
as collinite, tellinite, cutinite, alginite, resinite, sporinite,
micrinite, sclerotinite, semifusinite and fusinite, For ease of
usage, three maceral groups have been established-vitrinite, exinite
and inertinite. Each group includes macerals which, although not
identical, may appear similar under the microscope and in respect to
their technological properties.

Table 2- shows the terminology used in Stopes-Heerlen (S-H) system
of classification (16,17)

Flotation characteristics of group macerals and lithotypes are
different. If the surface properties of a coal's constituent are
well established, one would be able to predict the behaviour of a
particular coal.

Study of the maceral has shown that the fraction of the total
carbon in aromatic structures increases in the order : exinite,
vitrinite, inertinite. The hydroxyl group content, an indication
of polar character, decreases in the same order (14). One would

Table – 1   Approximate Values of Some Coal Properties in Different Rank Ranges   [11]

	Lignite	Subbit.	High Vol. Bit.			Bituminous		Anthra-cite
			C	B	A	Medium Vol.	Low Vol.	
% C (min. matter free)	65-72	72-76	76-78	78-80	80-87	80-85	85-90	90-95
% H	4.5	5	5.5	5.5	5.5	4.5	3.5	2.5
% O	30	18	13	10	10-4	3-4	3	2
% O as COOH	13-10	5-2	0	0	0	0	0	0
% O as OH	15-10	12-10	9	?	7-3	1-2	0-1	0
Aromatic C atoms % of total C	50	65	?	?	75	80-85	85-90	90-95
Av. no. benz. rings/layer	1-2	?		2-3			5?	>25?
Volatile matter, %	40-50	35-50	35-45	?	31-40	31-20	20-10	<10
Reflectance, %, Vitrinite	0.2-0.3	0.3-0.4	0.5	0.6	0.6-1.0	1.4	1.8	4
Density					increases			
Total surface area					minimum			
Plasticity and coke formation				only				
Calorific value, moist, Min. matter free, Btu/lb.	7,000	10,000	12,000	13,500	14,500	15,000	15,800	15,200

TABLE 2-Stopes-Heerlen System of Classification (16,17)

Macrolithotypes	Microlithotypes	Group Macerals	Macerals	Origin and Properties
VITRAIN	VITRITE (95 % vitrinite)	VITRINITE	Collinite Tellinite	Wood and cortical tissues. Either shows woody tissues clearly or is structureless. Grey clour by reflected light. Predominant constituent of most coals.
CLARAIN	CLARITE (Vitrinite dominent Exinite and Inertinite less Prominent)	EXINITE	Sporinite Resinite Cutinite Alginite	Micro and macrospore exines. Resines and waxes Cuticles Algea (not important except in Boghead coal) Except macrospore (often reddish in colour, exinite has grey colour, in general darker than vitrinite.
DURAIN	DURITE (Inertinite dominent Vitrinite and Exinite less prominent)	INERTINITE	Fusinite	Shows clear cellular structure.Cavities sometimes filled with mineral matter. Yellowish white by reflected light.
			Semifusinite	Wood and cortical tissues.The transition stage from fusinite to vitrinite white by reflected light.
FUSAIN	FUSITE (Inertinite)		Sclerotinite	Resinous and possibility fungal matter.Light grey to white.
			Micrinite	Occurs in massive or fine (granular) form.Structureless and opaque. White by reflected light.

expect that the order of hydrophobicity would increase in the order
exinite, vitrinite, inertinite, and this would be reflected in the
lithotypes in which they are a major component.This does not appear
to be the case,however,because the hydrophobicity of the lithotypes
usually increases in the order fusain, durain, clarain and vitrain
(14).One of the major factors for this order is due to the ordered graphite-like structure of inertinite, but mainly because of
the varying amounts of hydrophilic mineral matter intimately associated with inertinite.

2.2. Impurities associated with coal

Two types of ash forming components are encountered in the run-
of-mine coal : Fixed ash-Mineral inclusions in the coal seam which
is a term applied to the mineral matter associated with coal during
coalification. This is derived from the inorganic matter which grew
in the original coal-forming plants. Free ash-Carbonaceous shale,
clays, sandstone or other forms of rock, sulfur and deleterious
trace elements, immediately adjoin the coal seam, from hanging wall
and footwall, or as partings or layers of rock or clay within the
seam (intercalations) and are unavoidably extracted with it. Fixed
ash is defined theoretically the minimum ash achieved in a clean
coal.

Sulfur is present in coal in four forms: Elemental, sulfate,
organic and pyritic sulfur. Since the sulfate and elemental sulfur
are usually low, and the organic sulfur is intimately associated
with the coal, basically only pyritic sulfur may be removed by flotation or other physical separation methods provided the pyrite
particles are not too fine to be liberated.

The flotation properties of coals are very closely dependent
on the properties of inorganic impurities (e.g. if the amount of
clay minerals increase in coal, the effect on the flotation circuit
can be most deleterious). Classification of these impurities and
their properties affecting flotation are shown in Table 3 (18).

2.3. Natural Floatability of Coal

For flotation to occur, coal particles must become attached to
air bubbles passing through the suspension in a flotation cell and
be removed into the froth layer. The tendency for attachment is characterized by the wettability of the surface.

The atoms, ions or molecules which form solid bodies are interconnected by means of various bonds. When a crystal breaks down,some
unsaturated bonds arise at the surface. Solids, such as coal, consist of molecules which are held together by weak Van der Waals
bonds. When coals are fractured, these weak bonds are broken and

the surface is left in a relatively uncharged or non-polar condition; it is naturally hydrophobic, exhibiting a high contact angle.
On the other hand, solids generally have some sort of ionic or
metallic structure and on fracture their surface is left polar and
hydrophilic, with small or zero contact angle (14).

The complex, giant-molecular structure of coal undoubtedly results in some covalent bonds being broken, on size reduction, to
yield a surface which is largely neutral but with some degree of
polarization. This polarization is not sufficient to destroy the
floatability of coal but probably reduces it.

TABLE 3- Classification of Inorganic Impurities
Found in Coals (18)

Group	Representatives	Principal Properties Affecting Flotation
Sulfides	Pyrite, marcasite	Contaminate the concentrate with sulfur. Have high flotation activity, small grains can be floated with the same reagents as coal. Floatability deteriorates after oxidation.
Argillaceous substance	Kaoline, clays	Form large amounts of finely dispersed slimes which cause flotation to deteriorate.
Carbonaceous and bituminous shales	Carbonaceous shales, Bituminous shales.	Highly water-repellent, therefore pass into coal concentrates in flotation.
Highly soluble minerals	Gypsum, sylvite, potassium, sodium chlorides	Increase the pulp electrolyte concentration; which effect the separation in the process.
Non-Sulphides (silicates, carbonates, oxides)	Calcite, quarts, dolomite, feldspar.	Relatively low solubility, do not contain bitumens and humic acids. Water avid, they can be easily separated from coal by flotation.

The natural floatability of coal depends upon rank. As the
rank of coal decreases (i.e. its carbon content), it becomes more
difficult to treat by froth-flotation. Hence coking coals are most
suitable for flotation because good yields of relatively low-ash

84

product are obtained from them.

The essential condition for froth flotation is that a particle
of solid, wetted by water, should become attached to an air bubble
and to be carried to the surface. The extent to which the water is
displaced from the surface of the solid or in other words, the extent to which air/water, and solid/water interfaces are replaced
by an air/solid interface is characterized by the contact angle.
Brady and Gauger (19) evaluated the natural·floatability of coal
by using the contact angle. Brown plotted (14) also a curve for
the variation of contact angle with carbon content of coal. The
maximum contact angle was observed at 89 % C. Up to this value,
there is a loss of polar character by the decrease of hydroxyl
and acid groups so that the coal becomes more hydrophobic and
floatable, with a maximum floatability at 89 % C; above 89 % C the
hydrophobicity decreases slightly as the carbon atoms form structures with increasing order in three dimensions (14). Figure 1
shows the dependency of contact angle on rank (14).

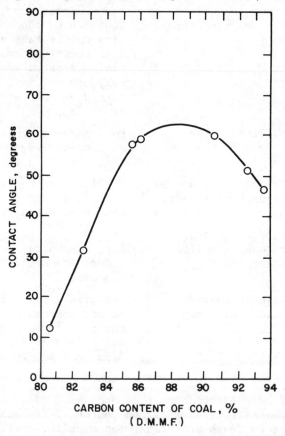

Figure - I   Variation of contact angle with carbon
content of coal [14]

The floatability of coal is also dependent on the ash content of coal. The floatability decreases as the ash content increases because the mineral matter associated with coal is hydrophilic.

The different constituents of banded coal have different natural floatabilities and contact angles. As stated earlier, ease of floatability usually increases in the order of fusain, durain, clarain, and vitrain which is also supported by the contact angle measurements (20) . The order found in a particular case depends upon the amount of mineral matter intimately associated with each macrolithotypes.

Table 4 shows the contact angles of a Turkish bituminous coal seam (Çay) and its lithotypes (21).

Table 4. Natural Contact Angles at Various Coal
Surfaces (21)

Coal Sample	Contact Angles (Degrees)[x]		
	Minimum	Maximum	Average
Çay Coal Seam[xx]	26	53	49
Vitrain	57	63	61
Fusain[xxx]	–	–	–
Durain	24	47	42

x-The average figure shown is the mean of 9 measurements taken
  at different parts of the surface.
xx-Turkish bituminous coal seam in Zonguldak Coal Basin.
xxx- Contact angle could not be measured.

Three theories have been suggested for the explanation of different floatabilities of various coals. Taggart et al (22) explained the difference in floatability between bituminous and anthracite coal by the difference in C/H ratio. Such an explanation is not applicable to the relative floatability of other coals and carbons.

Wilkins (23) indicated that the floatability of coals in - creased with increase of their carbon content or rank, however, bituminous coals that possess moderate carbon contents are actually more floatable than anthracite coals that have higher carbon contents.

Sun (24) developed a surface component theory which correlates bulk chemical analyses with the surface hydrophobicity. The gist of the theory is that coal consists of floatable and non-floatable chemical constituents and the floatability of the heterogenous solid is governed by the balance between the two groups of components. The surface of one coal predominantly composed of floatable components is more floatable than that of another coal whose surface is predominantly non-floatable components. The floatable constit - uents are considered oil-avid and water repellent, whereas the non-floatable components are water-avid and oil-repellent. This hypothesis takes into account of hydrocarbon molecules, excess carbon atoms, water molecules, and the relative number sulfur, oxygen, and nitrogen atoms. Mathematically, the hypothesis is expressed (in the condensed form) as follows :

$$F_c = x\left(\frac{H}{2.0796}\right) + y\left(\frac{C}{12} - \frac{H}{2.0796}\right) + 0.4\left(\frac{S}{32.06} - z\frac{M}{18}\right)$$

$$-3.4\left(\frac{O}{16}\right) - \left(\frac{N}{14}\right)$$

in which F_c is calculated floatability representing the balance between the floatable and the non-floatable surface components. The symbols H, C, S, N, O and M are respectively the weight per cent of hydrogen, carbon, sulfur, nitrogen, oxygen, and moisture of the ultimate analysis. The assumed numerical values of factors x, y and z are given in Table 5.

TABLE 5- The Assumed Numerical Values of Factors
x, y and z for Sun's Surface Component Theory (24)

Ash		Remnant Carbon		Moisture	
Ash % , as Received	x	Hydrogen % (dafb)	y	Moisture % as Received	z
0 to 8.9	3.5	0.08 to 0.28	1.2	0 to 14.1	4
9 to 13.0	3.0	0.29 to 1	0.8	14.1	3
13.0	2.5	1	0.6		

The calculated floatabilities of tested materials such as medium volatile bituminous coal, medium-volatile vitrain, high volatile bituminous coal, high volatile vitrain, fusain, and durain were found to be 10.64, 10.51, 9.81, 7.49, 5.94, 5.46 respectively(24)

Sun's surface component theory has two defects : i) Complete chemical analyses are rarely available, ii) since the analyses are based on the bulk sample, surface oxidation can completely destroy the estimate because surface oxidation leads to poor flotation.

2.4. Effect of Oxidation on Coal Floatability

There is a wide difference in the floatability of coals of different rank, and even of the same rank, depending on whether coals have been freshly mined or allowed to oxidize. Also differences in floatability, presumably due to oxidation, may occur within a particular seam because of ground water percolating through the coalbed or because of the proximity of the coal to the surface. Strip-mined coals are typically more difficult to float than deep-mined coals from the same seam (25, 26).

Generally, freshly-mined coal floats better than coal which has been exposed to the atmosphere. The effect is attributed to surface oxidation which is characterised by the formation of acid oxygen containing groups such as COOH, OH and the dissociation of these groups so as to reduce the hydrophobicity of the surface (27,28)

Probably all coals undergo atmospheric oxidation during mining and storage, and their degree of oxidation increases with temperature and the time of exposure(29,30). The resistivity of coals to oxidation increases with their rank, whilst the resistivity of petrographic components to oxidation increases in the order of vitrain, clarain, durain and fusain (20,25,31).

In coals of all ranks vitrinite is the most readily oxidized maceral, while exinite oxidizes less readily. Inertinite generally oxidizes only with difficulty (31).

The floatability of oxidized coal has been restored by dissolving the oxidized surface layer in a 1 % caustic soda solution, or by reduction with benzoyl alcohol. Flotation of oxidized coal can also be performed by using a cationic collector such as an amine(14,27,32).

3. OPERATIONAL FACTORS RELATED TO COAL

Major operational factors which affect the froth flotation of coal are as follows :
   i) Reagents
  ii) pH and zeta potential
 iii) Size and specific gravity of particles
  iv) Pulp density and temperature

88

   v) *Presence of slimes*
  vi) *Feed rate*
 vii) *Conditioning time*
viii) *Air flow rate*
  ix) *Type of flotation machine*
   x) *Flotation circuit*

3.1. <u>Reagents used in coal flotation</u>

As the coal is a cheap commodity, reagents including cheap
waste products obtained from coke-oven plants and petroleum distil-
lation are preferred. There are four types of reagents: nonpolar
oils, heteropolar reagents, inorganic electrolytes and depressants.

3.1.1. *Nonpolar oils*

The most effective collectors that can be employed in coal flotation are those oils with low solubility in water derived from
petroleum, wood, and coal tars. The more soluble and volatile compounds were found less active than the low volatile insoluble compounds (33).

Brown (14) states that fluid paraffinic hydrocarbon type
collectors, with as high a molecular weight as possible are probably the most effective collectors for coal. A practical limit of
twenty-carbon atoms in the molecule is suggested. Typical nonpolar
reagents, often termed oils, are the kerosenes (lamp oil), fuel
oil, crude petroleum, and certain coal-tar distillates. Nonpolar
reagents give practically no reaction with water dipoles and have
pronounced water-repellent properties due to the presence of weak
Van der Waals forces between the hydrocarbon molecules and covalent
bonds within the hydrocarbon chains. As nonpolar collectors have
no solidophil groups which are links between the radicals of the
reagent and the mineral surface in their structure, they do not
form oriented adsorption layers on the mineral surfaces. Their
attachment is based on adhesion (18) . Klassen (13) describes
the mechanism of oil attachment to coal surface as film formation. During film formation the reagents do not react chemically
with the mineral surface and their chemical composition does not
change.

For the formation of film, it is not sufficient for oil droplets to attach to the surface, they should spread over it. The
conditions under which spontaneous spreading of the droplets takes
place on the surface must be in the direction of reduction of the
free energy of the system (13). Light oils, i.e., those of low
molecular weight, spread easily over the surface of the coal, but
tend to be absorbed too quickly into its porous surface. Heavy
oils, on the other hand, require considerable agitation or conditioning to spread the droplets evenly over the surface,but they

ãre more effective as they stay on the surface, showing less
tendency to be adsorbed (34).

Oils added to the pulp are divided into droplets during conditioning process due to their low solubility in water. They may
also be added, already dispersed, in the form of an emulsion. It
has been shown that the preliminary emulsification of kerosene
improves the flotation and there is an optimum dispersion of the
emulsion corresponding to a droplet diameter of 5-10 microns(35).

Kerosene is one of the most widely used oily collectors in the
world. Like fuel oil, it is derived from petroleum and it is a
mixture of saturated liquid hydrocarbons. Such oily collectors do
not possess a constant composition, e.g. the composition of kerosene depends to a large extent on the source of the naphta from
which it is derived (13) . In the industrial application of kerosene in coal flotation, a frother should be added into the pulp.

Another widely used oily collector which is not derived from
petroleum is creosote. Industrially used creosote is a crude phe-
nolic mixture which is usually of coal-tar origin.

3.1.2. Heteropolar reagents

Heteropolar reagents have both polar and nonpolar parts in
their structure. In coal flotation, the most widely used heteropolar reagents are frothers. Frothers used in coal flotation may
also show collecting ability. In the flotation of coal,monohydric
aliphatic alcohols and monocarboxylic acids are good collectors
when the hydrocarbon radical contain between five to ten carbon
atoms, with an optimum of eight (13,36).

An excessive increase in the length of the hydrocarbon radical
reduces the flotation activity. Amongst the aromatic substances,
the most selective adsorption is obtained with molecules possessing
a hydroxyl polar group, while the least active polar groups are
those which contain the amino or sulfo groups (14) . Compounds
with a hydroxyl or carboxyl group also possess important collecting
properties, but their flotation activity decreases with an excessive
increase in the number of polar groups (37).

Evidence has been found recently regarding the bond between the
hydroxyl group and the surface of coal. The reagent is more stably
adsorbed that can be accounted for the adsorption in the electrical
double layer.This may be the result of an increased interaction
between the hydroxyl groups, due to hydrogen bonding (13) .
Infrared spectroscopic studies carried out with various alcohols
indicated the presence of H-bonds which supports this assumption
(38,39).

Heteropolar reagents make the coal surface more hydrophobic if
they are added at low concentrations. Further addition of reagent
possibly gives a second adsorbed layer on top of the first, with
the polar groups pointing away from the coal surface, and making
the surface less hydrophobic.

Amyl, hexyl, heptyl and octyl alcohols are examples of frothers.
Methyl isobutyl carbinol (MIBC) is also an aliphatic alcohol in
common use. Terpineol, pine oil and particularly cresylic acid
(a by-product of coke making) are especially attractive.

Coal generally responds readily to any of the frothers, but
the choice depends upon availability, price and effectiveness for
the particular coal being treated and its selectivity from gangue.
Chernosky (40) evaluated the coal flotation frothers on a "Yield-
Selectivity Cost" basis and found MIBC as superior to other reagents
tested.

For most coals, it is desirable to use both a frother and an
oily collector together. For example kerosene or fuel oil, with
an alcohol, cresylic acid, pine oil, or some waste scrub liquor
from coke plants. With cresylic acid and creosote, the most advantageous operating temperature is in the vicinity of $25^{0}C$ (41).

The application of ionic collectors such as xanthates, carbox-
ylic acids, sulfates and sulfonates in coal flotation is not more
than of a scientific interest. Long-chain amines have only found
application in the flotation of oxidized coals. Sun (27) and Wen
(42) found certain cationic collectors, laurylamine, rosin a-
mine-D acetate, and iso-amyl amine to be effective in the flotation of oxidized coals. This may be due to the fact that chemical
reaction and ionic attraction can take place between the positively
charged polar groups of amines and the negatively charged ions,
particularly carboxyl groups, of the oxidized surfaces of the coal.

3.1.3. Inorganic Electrolytes

It is possible to float many naturally hydrophobic minerals
in solutions of inorganic salts without additions of any other
reagents. Thus the first coal flotation practices were carried out
using sea-water in the mid-1920's (43). Nowadays, even differential flotations of relatively complex Pb-Zn and Cu-Pb-Zn ores
have been conducted successfully in sea-water media (44).

The floatability of naturally hydrophobic minerals in solutions containing only inorganic salts was explained first by the
frothing action of the aqueous solution of inorganic electrolytes.
According to Kovachev (45), at low concentrations, the electrolyte makes the mineral particles hydrophobic by decreasing the

surface potential and thus decreasing their hydration. At higher
concentrations, the electrolyte causes froth formation.

Coal has a considerable capacity for adsorption of electrolytes. Electrolytes change the electrostatic charge at the coal
surface by direct or exchange adsorption and render the surface
more or less hydrophobic by the consequent reduction or increase
of charge.

Electrolytes which are not adsorbed directly, may lead to compression of the double layer around the coal particles and so may
effect the stability of the hydrated layer at the surface (14).

Klassen et al (13) carried out a series of measurements of
contact angle, stability and the rate of attachment of particles
of coal to air bubbles, as well as actual flotation experiments
using sodium chloride solutions. Results showed that additions of
sodium chloride improved the attachment of coal particles to air
bubbles.

The use of inorganic electrolytes in flotation, sometimes
known as "salt flotation", was not widely developed because of
their corrosive properties, but today shortages of fresh water or
the poor qualities of some so-called fresh waters have forced several mills to conduct flotation in sea-water (44). Corrosion
problems have been overcome partly by resistant linings or by using
more resistant metals. It is interesting to note that the use of
sea-water yields a drier froth in the concentrate with a water content of 30 to 35 percent while in flotation with nonpolar reagents,
the water content is around 50 to 60 percent (14). Salts also
increase the settling rate of minerals in the froth concentrate
and permit drainage, no filtration may even be needed (44) .

Salt flotation is very rapid and selective. Kharlamov (46)
indicated that the advantages of salt flotation were apparent in
the first few minutes of flotation. It was also found that the flotation time is 30 percent shorter than in fresh water (47) The
use of salts also caused production of concentrates of comparatively low ash and sulphur content in the coal (46).

Combined addition of inorganic electrolytes and nonpolar reagents are especially important in coal flotation practice. The
addition of a very small amount of kerosene sharply increased both
the rate of flotation and the recovery (45) . Kovachev et al
(47), found that the identical flotation activity was obtained
in sea water with only 1/3 to 1/5 of the amount of nonpolar collector used in fresh water.

3.1.4. Depressants

Non-electrolytes and salts of inorganic acids are generally used as depressants in coal flotation. The former will depress coal while the latter are more effective on gangue minerals. Two major types comprise the non-electrolyte group of regulation agents, carbohydrates and tannins. The carbohydrates, dextrin, starch etc. are high molecular weight compounds containing a large number of strongly hydrated polar groups (OH, COOH, CO, COH, etc). The second type of non-electrolyte, tannins are composed of complex molecules with phenols and thio sulphates (mainly sulpholignin acids). The common tannin in use is quebracho.

Two major salts of inorganic acids used in coal flotation are sodium silicate and sodium cyanide. The latter salt is used for depressing pyrite while the former is used to disperse clays and shale.

3.2. pH and Zeta-Potential

All the coal samples (weathered or unweathered) and their lithotypes have negative zeta-potentials in distilled water(21, 26,48,49).

It is appropriate to discuss the surface negativity of coal in respect to its nature. Coal is a heterogenous substance and its surface is anistopropic in nature. As it is described by Chander, Wie and Fuerstenau (50) , the anisotropic surface consists of two parts, one which is formed by the rupture of Van der Waals bonds and is hydrophobic and the other which is formed by the rupture of ionic or covalent bonds and is hydrophilic. Kitchener(51) assumes that pure graphite-like areas on the surface of coal are hydrophobic and practically non-ionogenic (inert) and that there are also hydrophilic sites on the surface which are polar,probably "Oxide complexes", which behave as weakly acidic groups, at the edges and places of lattice breakage. Besides these, coal contains some inorganic impurities which are hydrophilic in nature. Therefore, the coal surface will show negative charge due to the presence of hydrophilic sites at the surface.

The variation of zeta-potential of a Turkish bituminous coal seam (Çay) and its lithotypes with pH is shown in Figure 2 (21). Figure 2 shows that when the hydroxyl ions are added into the pulp, the coal surface becomes more negative due to the adsorption of hydroxyl ions. Conversely, when the pH of the solution in contact with the coal surface is lowered, hydronium ions are adsorbed until the negative charge is reduced to zero. As the acidity of the solution is increased, the surface acquires a positive charge. Thus, various concentrations of hydronium and hydroxyl ions not only change the magnitude of zeta-potential, but also the sign of it.

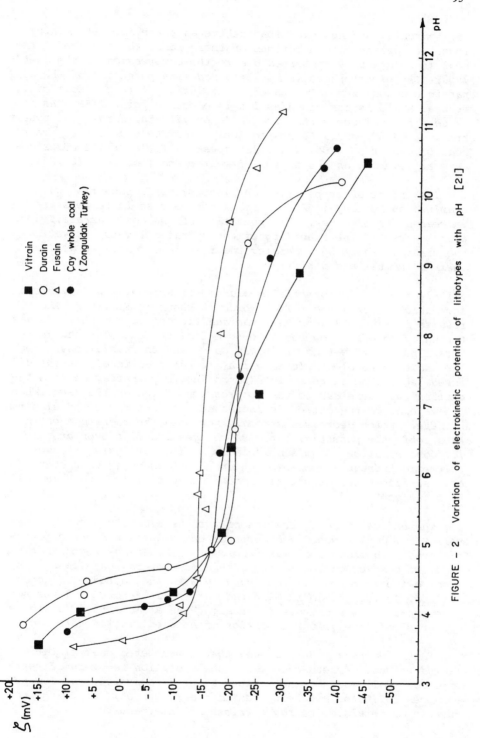

FIGURE – 2   Variation of electrokinetic potential of lithotypes with pH [21]

So, hydronium and hydroxyl ions behave as potential determining ions. At certain concentrations of these ions, the surface of the coal in solution is uncharged due to the compression of the double layer. The electrophoretic mobility and zeta-potential of the coal particle become zero. The pH of the solution at this electrically neutral state is described as its isoelectric point (IEP). As it is seen from Figure 2 that the IEP's of vitrain, durain and fusain are 4.10, 4.50, and 3.60 respectively. The IEP's of their parent coal occurs at pH 4.05. Generally speaking IEP's of all coals occurs at lower pH as the degree of weathering increases (21,26).

The effect of temperature on electrokinetic potential was examined by Kovachev (52). He found that a rise in temperature increased the electrokinetic potential of the coal surface.Temperature affected the stability of the hydrated layers, solvating the mineral grains by changing their surface charge as well as their structure and mobility.

Electrokinetic potential measurements were made with various inorganic electrolytes (21) which is shown in Figure 3. The results in Figure 3 show that the addition of sodium chloride has virtually no effect on the zate-potential of coal. The charge reversal point did not occur in Na^+, Ca^{++} , and Zn^{++} solutions. They are indifferent electrolytes which are adsorbed merely in the diffused layer. The decrease in electrokinetic potential and the hydration layer are due to the compression of the double layer.When the valence of the cation is increased, as in the case of aluminum sulfate, cation decreases strongly not only the magnitude of the electrokinetic potential but also changes its sign even at very low concentration. Aluminum ion acts by its hyrolysis, producing polymeric hydroxy - complexes which adsorb on negatively charged coal surface by electrostatic force, converting it into "alumina-like surface" (58).

The effect of concentration of various alcohols such as butanol, MIBC, amyl alcohol and heptanol on the electrokinetic potentials of bituminous coal was studied (21). It was found that heptanol with seven carbon length of hydrocarbon radical lowers the electrokinetic potential continuously with the increase of concentration. As the change in magnitude of the zeta-potential was not accompanied by the change of its sign, the attachment of alcohols concerns only the diffused portion ofthe electrical double layer.

Yarar and Leja (26) showed that unweathered coal flotation is independent of zeta-potential, the flotation recovery of weathered coal follows the same pattern as zeta-potential, and low pH enhances collector uptake by deslimed surface coal due to the reduction of repulsive energy barriers at interfaces.

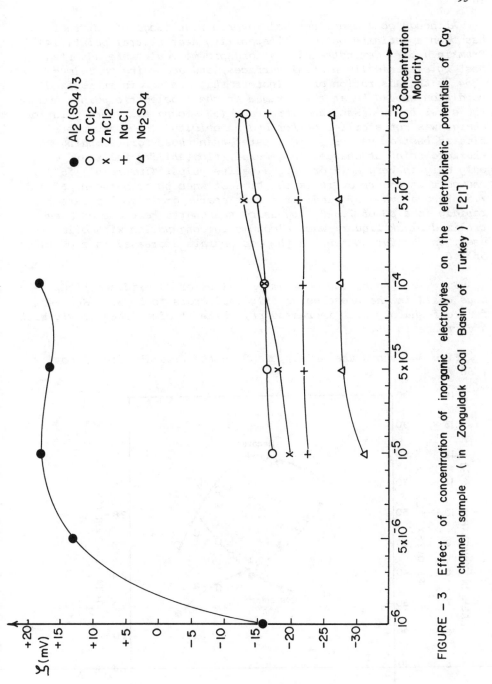

FIGURE – 3   Effect of concentration of inorganic electrolytes on the electrokinetic potentials of Çay channel sample (in Zonguldak Coal Basin of Turkey) [21]

Although coal can float well over a wide range of pH, there has been a definite optimum floatability near neutral point(14) Frumkin (53) demonstrated that organic molecules are adsorbed best by electrically neutral surfaces, and according to Talmud (54), the adsorption of collectors that are not in an ionized condition should be at the maximum at the isoelectric point. On the other hand, Jessop and Stretton (55) found that zero point of charge was not a criterion for hydrophobicity. The coals tested floated best at values of pH greater than those corresponding to the zero point of charge. In general, floatability falls off mark- edly only in very acid or very alkaline pulps. Zimmerman (56) , showed that the recovery is the highest when pH is between 6 and 7.5. Konovalova (57) indicated that coarse grain coal floats more rapidly in a pH of 7.8-8.0 by using sulfonated kerosene and neu- tralized black liquor; when alcohols in conjunction with oils are used, the sulfur content in the concentrate decreased in a pH range of 5.0-9.0.

Sun (27) found the optimum pH value for unoxidized bitumi - nous coals in the presence of oily collectors to be 7.5. Recovery fell off sharply with increased pH, while it diminished slowly with a decrease in pH.

Figure 4 shows the effect of pH on the floatability of coal (56).

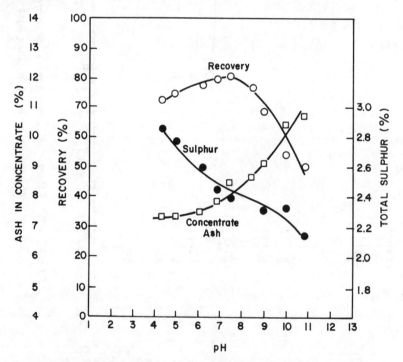

Figure - 4   Effect   of   pH   on   the   floatability [56]

3.3. Size and Specific Gravity of Particles

The organic component of coal is of lower specific gravity than its impurities, and the upper size limit in coal flotation is higher than in mineral flotation. The maximum size of bituminous coal particles (sp. gr. 1.35) floated in a laboratory cell was 6.7x3.3 mm. The maximum size of anthracite particles (sp.gr. 1.54) floated by 1.17x0.83 mm. Here the bituminous coal has not only a higher natural floatability than the anthracite but also a lower specific gravity (59).

Coarse coal particles, up to 1.7 mm diameter, have been concentrated by a flotation -agglomeration process. The conditions that favours coarse coal flotation are often detrimental to the rejection of ash and pyritic-sulfur in the fine coal fraction.

Large quantities of collector is necessary for the flotation of coarse particles and it is not always possible to use collectors in these quantities because of reduced selectivity as well as the inherent cost factor. For these reasons, coal particles larger than 28 mesh are better treated by other methods than flotation.

Some problems may also be encountered because of the wide range of particle size in the feed to flotation units. For instance when 0.84x0 mm at 15 percent solids feeds to flotation units, the fines float rapidly so that only coarse coal is left in the later stages of the circuit. As a result, dilution is low and reagent concentration too weak for effective recovery of the coarse fraction. Even with stage addition of reagents, best results are not always obtained on this size range of feed. For this reason, it sometimes may be advantageous to separate the feed at approximately 48 mesh (0.297 mm). Ideal size for coal flotation occurs within the range of 48 to 150 mesh. Figure 5 shows the floatability of coal as a function of particle size (60).

3.4. Pulp density and temperature

Metalliferous ores and silicate minerals have been found to respond best to the flotation process at a pulp density of approximately 20 to 30 percent solids. Coal, by reason of its low specific gravity and friable tendencies, occupies from two to three times more space for the same tonnage at the same pulp dilutions as heavier minerals. Hence to achieve the same looseness of pulp in dispersed condition, a coal pulp must carry a smaller percentage of solids. Pulp density varies in practice from 6 to 25 percent, depending on the rank, type, ash content and size distribution of the coal feed. If a very low ash product is required, results are generally better at the lower concentrations. Where

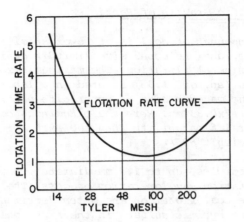

Figure – 5 Floatability as a function of
particle size [60]

the feed is mostly minus 200 mesh, a dilute pulp is required (56,
61). The coarser the feed, the higher the pulp density required
for most effective results. For an average condition, treating
a minus 48 mesh feed containing perhaps not more than 25 percent
minus 200 mesh particles, Zimmerman (61) has found a feed density
of 10 to 14 percent as being desirable, raising it to 14-18 percent with an increase in coarse sizes; and lowering it to perhaps
5 percent in treating an all slime (minus 200 mesh) feed.

Generally, a change in temperature affects the reagent viscosity and interfacial tensions as well as their collecting and
frothing properties.In coal flotation, flotation recoveries were
markedly affected by temperature variations when certain reagents,
particularly simple aliphatic alcohols, were used. For other re-
gents, particularly oily types, flotation recoveries were substantially independent of variations in pulp temperature (62).

3.5. Presence of Slime

Kaolinite is the commonest of the clay minerals associated
with coal. The presence of finely divided clay has been found to
reduce the recovery of coal by flotation and to increase the reagent consumption. This is due to the coating of the coal particles
with a very thin layer of clay which renders them hydrophilic
rather than hydrophobic,thus reducing the probability of adhesion
between the air bubble and coal particle. Slime coating also
makes difficult the subsequent filtration of the concentrate.
Brown (63) found that when the influence of kaolinite was investigated, it was only with the continuous technique that an accurate

forecast of the behaviour under commercial conditions could be
made. With the finest sizes, continuous testing showed that the
falling off in the efficiency of flotation was around 50 percent.

3.6. Feed rate

Because of the low specific gravity of coal and the large
number of particles per unit weight, it is absolutely necessary
that the rate of feed to the cells be regulated within as close
limits as is possible . High variations in percentage of solids
or in tons per hour of solids throws any recirculating circuit
entirely out of balance and results in poor concentrates, poor
tailings, and a heavy build up of recirculating middlings.

3.7. Conditioning time

The conditioning time in coal flotation depends especially
on the type of reagents used and the rank of coal. The cresylic
creosote-based reagents (which is the frother/collector commonly
used in coal flotation plants) must be thoroughly dispersed
throughout the pulp so that all the solids are brought into contact
with the oil. With coals whose oil consumption is moderate or high,
adequate conditioning improves yield and ash content in froth as
well as reducing oil consumption. The average conditioning time
for coal in practice varies within the range of 2 to 5 minutes.
While preconditioning may be dispense with for readily floatable
coal, it is a prime necessity for the difficult coal and the period is usually longer.

Plants are in operation where this conditioning takes place
in the first flotation cell, but it is generally more satisfactory to have a separate conditioner tank.

The rate of reagent addition must be controlled and a simple
reagent feeder may be mounted above the conditioner tank or the
feed launder. In certain cases multi-stage reagent addition is
found to be necessary and a series of reagent addition point may
feed directly into a number of cells.

3.8 Air flow rate

Air flow rate should be as high as possible without "burping".
Increasing the air rate through 0.3 to 2.0 scfm would significantly increase the yield/ash recovery ratios (64). Superior froth
products in respect to yield and ash would be obtained at the
2 scfm level of aeration. In cells that draw their own air, this

is not a variable which can be controlled independently of speed.

3.9. Type of Flotation Machines

The equipment used for coal flotation can be categorized into two main types-mechanical and pneumatic (cyclo-cell). Table 6 shows the major flotation machines used in coal flotation.

Table - 6   Major   Flotation   Machines   [65]

Manufacturer	Open-Flow-Thru Standard Cell Size m3	Cell-to-Cell Standard Cell Size m3
Agitair (mechanical)	1.7, 2.8, 4.2, 5.7, 8.5, 11.3, 14.2, 28.3	
Cyclo-Cell (pneumatic)	6.8 through 22.7	
Denver (mechanical)	1.4, 2.8, 8.5, 14.2, 36.1	1.4, 2.8, 5.7, 8.5, 11.3, 14.2
Humboldt (mechanical)	2.8, 5.7, 11.3	
Minemet (mechanical)	4.2, 8.5, 14.2	
Nagahm (mechanical)	1.7, 8.5	
Outokumpu OY (mechanical)	2.8, 1.7, 36.8	
Sala (mechanical)	2.8 5.7 11.3	
Unifloc (mechanical)		1.4, 2.8, 5.7, 8.5, 11.3, 14.2
Wemco (mechanical)	1.7, 2.8, 4.2, 8.5, 14.2, 28.3	

Mechanical flotation machines can be used as single cells, multiple cells in series, in parallel or with multiple rotors or agitators in a single trough with no partitions between them. The two major tank designs are used in coal flotation : The open-flow thru design in which the major flow of pulp is directed through the bank of cells, with recirculation occurring within each individual cell. Cell-to-cell design in which the major flow of pulp must pass over a walled weir between each cell before passing to the next cell. In Europe the cell-to-cell flotation design is preferred for deep cleaning of metallurgical grade coal because of a reported higher selectivity. Open-flow-thru design is commonly used in USA because of higher throughput.

Unlike the mechanical agitation cells cyclo-cell consist of a tank or trough with two or more vortex chambers in series. Agitation and aeration are supplied by pumping the pulp and air tangentially, as in a cyclone, but using a submerged vortex chamber and no cone. Submerged vortex chamber discharges a high velocity jet of agitation water and air in the form of a hollow cone. Air is released in the center of this cyclonic cone.

The pneumatic cells are especially suitable for the flotation of friable coals and they have large capacities.

3.10. Flotation Circuit

"Practically anything can be floated by the proper use of reagents and circuit design". This was the remark made by Zimmerman for stressing the importance of both reagents and flotation circuit in flotation operation (61).

There are as many variations of coal flotation circuits as there are coal seams. However, it is imperative that a good circuit should give a product low in both ash and sulfur contents with a high recovery. Table 7 shows various kinds of circuits with their products quality (65).

Table – 7   Coal   Flotation   Circuits   [65]

FLOTATION CIRCUIT		YIELD % by wt	PRODUCT ash adb	TAILINGS ash adb	E
Single Stage Flotation	(A)	59.6	11.6	27.0	140
Desliming	(B)	60.0	8.2	31.3	230
Separate Conditioning of the Coarse and Fine Size Fractions	(C)	51.0	13.0	23.5	90
Two Stage Conditioning	(D)	65.0	12.8	28.6	145
Split Feed Flotation	(E)	75.2	11.2	39.5	265
Two Stage Reagent Addition	(F)	79.6	11.0	43.6	315
Reflotation of Classified Tailings	(G)	81.0	9.9	58.0	475

Two-stage reagent addition circuit (circuit F) has been used in a number of coal washeries in Europe and the United Kingdom for the improved collection of the coarse size fractions. This model for this circuit consisted of adding 1 kg/ton of reagent, performing a flotation and then repeating the procedure on the tailings from the first flotation.

4- COST OF COAL PREPARATION PLANT

Zimmerman(3)has classified the coal preparation processes

into five levels and he estimated their approximate capital and operating costs for a standard coal preparation plant which treats all sizes down to zero, consisting of coarse and fine coal cleaning circuits including froth flotation and using the best circuits for optimum results for moderate and difficult coals, the capital cost was given as 25 million U.S. dollars and operating cost as 1.00 U.S. dollar per ton of R.O.M. coal treated. The calculations were based upon a 1000 TPH plant which treats 3 million tons of R.O.M. coal annually.

5. CONCLUSION

Fine coal cleaning has become an important aspect of coal preparation due to the increased production of fine coal by the use of mechanical mining methods and by the need to reduce sulphur and ash in clean coal products. Froth flotation is a widely used cleaning process for fine coal. There are many variations in coal characteristics that affect its flotation behaviour. A successful coal flotation plant operation is only possible through the optimization of a number of process variables in order to obtain high-percent recovery and high quality coal products. Further research is required to understand the factors influencing the coal flotation process.

ACKNOWLEDGEMENTS

The author thanks the various authors and organizations for permission to reproduce figures, tables and results.

REFERENCES

1. Gaudin, A.M., Flotation, 2nd. ed., McGraw-Hill Book Company, Inc., 1957, New York
2. Gibbs and Hill, Inc., Coal Preparation for Combustion and Conversion, AF-791 Research Project 466-1, Final Report, May 1978, Prepared for EPRI, 3412 Hillview Avenue, Palo Alto, California 94304, p. 2-8
3. Zimmerman, R.E. New Trends in Coal Preparation, Course in Coal Technology, Oct-Nov. 1982, Maracaibo, Venezuela, Lecture No: 1,2,3,4
4. Aplan, F.F, Coal Flotation, Flotation, A.M. Gaudin Memorial Volume, M.C. Fuerstenau (Ed), AIME, 1976, New York, p.1235-1264.
5. Blagov, I.S., Coal Preparation in the Soviet Union,World Coal, May 1979, p.26-30
6. Özbayoğlu, G. Coal Preparation in Turkey, World Coal, Febr. 1980, vol.6, Nm 2, p.22-24
7. O'Brien E., Fine Coal Cleaning in the 28x0 Fraction, Mining

Engineering, August 1980, p.1213-1214

8. Kessler, M.F, Interpretation of the Chemical Composition of Bituminous Coal Macerals, Fuel, vol.52, July 1973, p.191-197

9. Goldman, G.K., Liquid Fuels from Coal, Chemical Process Review No.57, Noyes Data Corporation, 1972, New Jersey

10. Hill, G.R., L.B.Lyon, A new Chemical Structure for Coal,Ind. Eng. Chem., 1962, 54, No. 6, p.36

11. Given, P.H, Short Course on Coal Characteristics and, Coal Conversion Processes, The Pennsylvania State University, May 19-23, 1975.

12. Lowry, H.H., Chemistry of Coal Utilization, Supp. volume, John Wiley and Sons, 1963, N.Y.

13. Klassen, V.I., V.A. Mokrousov, An Introduction to the Theory of Flotation, Translated by Y.Leja and G.W. Poling, Butterworths, 1963, London

14. Brown D.J. Coal Flotation, Froth Flotation. 50th Anniversary Volume, D.W. Fuerstenau (Ed), AIME, 1962, N.Y. p.518-538

15. Stopes, M.C.,On the Four Visible Ingredients in Banded Bitu - minous Coal ; Studies in the composition of coal, No.1, Proc. Royal Society, London, 1919,90,p.470-487

16. Leonard, J.W., D.R. Mitchell., Coal Preparation, AIME, N.Y. 1968, 3rd. Ed., p.1-10

17. Smith, A.H.V., Petrographic Analysis of coal by reflected light, N.C.B., Coal Survey North Eastern Division, Ref. S/ 2447, 1958, June

18. Glembotskii, V.A., V.I. Klassen, I.N. Plaksin, Flotation, Translated by R.E. Hammond., Primary Sources, N.Y.1972,p.399

19. Brady, G.A., A.W. Gauger, Properties of coal surfaces, Industrial and Eng. Chemistry, Vol. 32, 1940, p.1599-1604

20. Horsley, R.M., H.G.Smith, Principles of Coal Flotation, Fuel, London, 1951, Vol.30,p.54

21. Özbayoğlu, G. "Determination of the Flotation Characteristics of several Turkish Bituminous Coal Seams in Zonguldak Coal Basin, Ph.D. Thesis, 1977, Middle East Technical University, Ankara-Turkey

22. Taggart, A.F., G.R. Del Guidice, A.M. Sadler, M.D.Hassialis, Oil-Air Separation of Non-sulfide and Non-metal Minerals, Trans. AIME, 134, 1939 p.180

23. Wilkins, E.T., Coal Preparation, Some Development to Pulverized Practice, Conference on Pulverized Fuel Harrogate, England, 1947, p.398

24. Sun, S.C, Hypothesis for Different Floatabilities of Coals, Carbons and Hydrocarbon Minerals, Trans. AIME, Min. Eng. Jan. 1954, p.67-75

25. Miller, K.J., Flotation Study of Refractory Coals, U.S.Bureau of Mines, 1977, RI 8224

26. Yarar, B. J. Leja, Flotation of Weathered Coal Fines from Western Canada, IX Intern. Coal Preparation Congress, New Delhi, 1982, Paper C-5

27. Sun, S.C., Effects of Oxidation of Coals on Their Flotation
    Properties, Mining Engineering, April 1954, p. 396-401
28. Baranov, L.A., F.M. Stankevich, The Effect of the Surface
    Charge of Oxidized Coal on Its Floatability , Soviet Mining
    Science, Int. Journal of Mineral Processing, 1971, vol.7,p.101
29. Iskra, J., J. Laskowski, New Possibilities for Investigating
    Air-oxidation of Coal Surfaces at Low Temperatures Fuel,1967,
    vol.46, p. 5-12
30. Gayle, J.B., W.H. Eddy, R.Q. Shotts, Laboratory Investigation
    of the Effect of Oxidation on Coal Flotation, U.S. Bureau of
    Mines, 1965, RI 6620
31. International Handbook of Coal Petrography, 2nd. Edition,
    1963 (English) International Commission for Coal Petrology,
    Centre National de la Recherche Scientifique, Paris.
32. Nimerick, K.H., B.E. Scott, New Method of Oxidized Coal
    Flotation, Mining Congress Journal, Sept. 1980, p.21-23
33. Taubman, A.B., L.P. Yanova, Peculiar Mechanism of Action of
    Nonpolar Flotation Reagents on the Flotation of Coal, CA 57,
    1199d
34. Dell, C.C., The Principles of Froth Flotation of Coal,Miner-
    als Engineering Handbook, June 1976, p.43
35. Kovachev. K.P., A. Venev, Vuglishta, Emulsification of the
    Reagents in Coal Flotation, CA 61, 14415e.
36. Mainhood, J,,P.F. Whelan, Flotation Frothers for Low-Rank
    Coals, J. Applied Chemistry, March 1955, p.133-143
37. Whelan, P.F., Froth Flotation Reagents for Coal, Distilla-
    tion Fractions of Commercial Oils and some simple Phenolic
    Compounds, J. Applied Chemistry, July 1953, p.289-301
38. Plaksin, I.N., V.I. Solnyshkin, Infrared Spectra of some
    Flotation Agents, CA 55, 6140c
39. Zubkova, Yu, N., R.V. Kucher, A.V. Rukosueva, Adsorption of
    Methanol and Heptanol by Donetz Coals in Various Stages of
    Metamorphism as Studies by Infrared Spectroscopy, CA.70 ,
    98538z
40. Chernosky, F.J., Evaluation of Coal-Flotation Frothers on a
    Yield-Selectivity—Cost Basis, Trans. of Soc. of Min. Eng.,
    March.1963, p.24-27
41. Bailey, R., P.F.Whelan, The Influence of Pulp Temperature on
    the Froth Flotation of Four British Fine Coals, Journal Inst.,
    Fuel, 25,1953, p.304-307
42. Wen, W.W., S.C. Sun, An Electrokinetic Study on the Amine
    Flotation of Oxidized Coal, SME-AIME Fall Meeting and Exhibit,
    Denver, Colorado, Sept. 1976, Preprint No. 76F 343.
43. Minerals Separation Ltd., German Patent No.478065, 1920
44. Smillie, L.D.H., Sea-water Flotation, Can. Min.J., vol.6,
    June 1974, p. 68-70
45. Kovachev, K.P., The Mechanism of Action of Inorganic Electro-
    lytes in the Flotation of Non-Polar Minerals. CA 56,7613a
46. Kharlamov, V.S., Desulfurization of Coal by Salt Flotation

CA 57, 1202g.

47. *Kovachev, K.P., S. Georgieva, Flotation of Low-Grade Bitumi-nous Coals in Sea Water, CA 72, 14488p .*

48. *Campbell, J.A.L., S.C., Sun Bituminous Coal Electrokinetics, Soc. of Min. Eng., AIME Trans Vol. 247, 1970, p.111-114*

49. *Campbell, J.A.L., S.C.Sun, Anthracite Coal Electrokinetics, Soc. of Min. Eng., AIME, Trans. vol.247, 1970 p.120-122*

50. *Chander, S., J.M.Wie, D.W. Fuerstenau, On the Native Floata-bility and Surface Properties of Naturally Hydrophobic Solids, Advanced in Interfacial Phenomena of Particulate/Solution/Gas Systems, Applications to Flotation Research, P.Somasundaran, R.B. Grieves (Ed), 1975, AIChE, Symposium Series No.150, Vol.71, N.Y.*

51. *Kitchener, J.A., Written Communication in April, 1975*

52. *Kovachev, K.P., Temperature D.ependence of Non-Polar Mineral Flotation with Inorganic Electrolytes, Tekhnika (Sofia),1961 9, p.16-19.*

53. *Frumkin, A.N., Z. Physik, 35, 1926, p. 792*

54. *Talmud, D., N.M. Lubman, Z.Kolloid, 50, 1930, p.159-163*

55. *Jessop, R.R., J.L.Stretton, Electrokinetic Measurements on Coal and a Criterion for its Hydrophobicity, Fuel, 1969, vol.48, p.317-320*

56. *Zimmerman, R.E., Flotation of Bituminous Coal, Trans. AIME, vol.177, 1948, p.338-356*

57. *Konovalova, T.F., Effect of pH of the pulp on the Floatability of Coal, CA 55, 23973 b*

58. *Mackenzie, J.M.W., Zeta-potential Studies in Mineral Process-ing, Measurement Technique and Applications, Minerals Sci. Eng., July 1971, p.25-43*

59. *Sun, S.C, R.E. Zimmerman, The mechanism of Coarse Coal and Mineral Froth Flotation. Trans. AIME, vol.187, 1950, p.616-622*

60. *Zimmerman, R.E., Flotation Reagents, Coal Preparation, J.W. Leonard, D.R. Mitchell (Editor), 3rd. ed. AIME, 1968, N.Y. p.10-76.*

61. *Zimmerman, R.E., Froth Flotation in Modern Coal Preparation Plants, Mining Congress Journal, May 1964, p.26-32*

62. *Gayle, J.B., A.G. Smelley, Effects of Temperature Variations of Contact Angles for coal and related substances, U.S.Bureau of Mines, RI5585, 1960*

63. *Brown, D.J, H.G. Smith, Continuous Testing of Frothers,Colliery Engineering, June 1954, p.245-250.*

64. *Cavallaro, J.A., Operating variables in Coal Flotation,Mining Congress Journal, Sept. 1970, p.49-57*

65. *Plouf, T.M., Froth Flotation Techniques Reduce Sulfur and Ash, Mining Engineering, August 1982, p. 1218-1223.*

GRAVITY CONCENTRATION METHODS

R. O. Burt

Tantalum Mining Corporation of Canada Limited

ABSTRACT

Gravity Concentration - the separation of minerals by virtue of specific gravity - is one of the oldest forms of mineral processing. Whilst its relative importance has declined in the twentieth century, the high cost of alternative processes, along with the development of a range of high capacity devices has led to something of a renaissance of gravity concentration.

Even with its long history the mechanisms of gravity concentration are imperfectly understood, which is partially the reason that the range of equipment available to the gravity concentration engineer is larger than all other branches of mineral processing put together. This paper briefly introduces the more important of these devices, comparing their relative uses.

Development of the gravity concentration flowsheet is discussed and the paper finishes with the description of two plants, one treating iron ore and the other cassiterite, as an indication of the diversity of gravity concentration flowsheets.

INTRODUCTION

Gravity Concentration is, next to hand picking, the oldest form of mineral processing; and it was the dominant process used for over 2000 years (1, 2). It is only in the twentieth century that its importance has declined, with the development of such processes as flotation, magnetic separation and leaching.

Nevertheless, Gravity Concentration is not dead - far from it. As recently as 1978 it was reported (3) that the total mineral tonnage treated by this process in the United States was greater than that processed by flotation. Coal represents the bulk of the ore treated by gravity separation, with iron ore representing a major portion of the balance.

It would, however, be erroneous to assume that gravity separation is applicable only to coal and iron ore separation and a few obscure separations where flotation has failed. In general, gravity separation has a lower installed cost per tonne of throughput than flotation for any given job, and usually has a lower installed power requirement. Gravity separation does not use expensive reagents, the cost of which (for flotation) is continually spiralling upward. With the exception of slime disposal (common to flotation), the environmental impact of gravity plant effluent is considerably less than that for flotation due to the absence of organic chemicals and their reaction products (4).

Modern changes in the gravity concentration field are mainly related to the development of a range of high capacity, high efficiency, but inexpensive equipment: modern gravity concentration plants are relatively simple and inexpensive compared to those in use previously.

Indeed, such is the relative cheapness of coarse gravity concentration there is benefit in its consideration for the rejection of barren waste at a relatively coarse size even in large flotation plants.

Gravity concentration is used today for the treatment, not of one or two minerals, but for a diverse range; from andalusite to zircon, from coal to diamonds, from mineral sands to metal oxides and from industrial minerals to precious metals. A partial list of minerals treated by gravity concentration is given in Table 1.

TABLE 1

Minerals Recovered by Gravity Concentration

Coal	Barite	Gold
Uranium	Fluorite	Silver
	Andalusite	Platinum
Iron Ore	Cassiterite	
Minerals Sands	Wolframite	Diamonds
Chromite	Scheelite	
Manganese	Tantalite	Sulphides

PRINCIPLES

Gravity Concentration may be defined as the separation of two or more minerals, usually of different specific gravity, by their relative movement in response to the force of gravity and one or more other forces, one of which is generally the resistance to motion by a viscous fluid such as water.

Factors which are important in determining the relative movement include the weight, size, shape, and specific gravity of the particle, not only in absolute terms, but relative to all others particles. If, in a hypothetical two mineral separation, any one of these factors is significantly different, but all others are the same, separation is relatively easy. In nearly every case, however, there will be a range of different minerals, each of different specific gravity, and with a range of particle weight, size and shape: the ease or difficulty of separation of one species from another will depend on the relative differences in these factors, and whether such differences assist or oppose separation.

Some idea of the amenability of separation of two minerals can be obtained from the concentration criterion, which is usually defined as: The specific gravity of the heavy species minus the density of the suspending fluid, divided by the specific gravity of the light species minus the density of the suspending fluid. Algebraically this is:

$$\text{Concentration Criterion (CC)} = (\sigma_h - \sigma_f)/(\sigma_l - \sigma_f) \qquad (1.)$$

where:

σ_h = specific gravity of the heavy mineral-kg m^{-3} x 10^{-3}
σ_l = specific gravity of the light mineral-kg m^{-3} x 10^{-3}
σ_f = density of suspending fluid-kg L^{-1}

In order to allow for differences in particle shape, the Concentration Criterion must be multiplied by a shape ratio factor. This factor is the quotient of the shape settling factor for the heavy mineral and the shape settling factor for the light mineral. The shape settling factor is: the ratio of terminal velocities of two particles of the same mineral, of the same measured particle size, but different shape; the first particle to be that for which the ratio is required, and the second a sphere. Provided particle shape is taken into account, the Concentration Criterion can be quite useful. If shape is ignored, the engineer may well be in for some unwelcome surprises.

The Concentration Criterion is usually compared, at the correct particle size, with a standard curve. Such a curve is

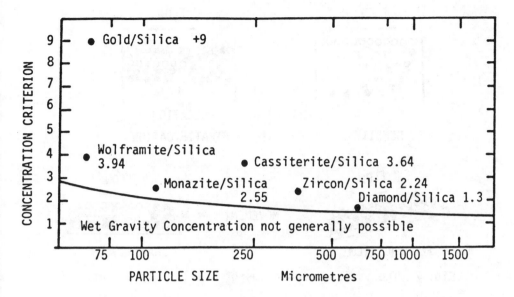

Fig. 1   The Concentration Criterion at Different Particle Size

shown in Fig. 1.   The curve itself indicates the point at which gravity separation becomes virtually impossible (5).   The further above the curve the determined Concentration Criterion is, the easier should be the separation.

Mechanisms of Concentration

A review of the available literature indicates that there is no one mechanism that can satisfactorily explain the behaviour of any particular concentrator and that a combination of two or more mechanisms comes closer to (but rarely completely) explaining the operation of the separator.   Conversely however, various mechanisms that have been proposed are common to apparently diverse items of equipment and gravity concentration processes (6).

The various mechanisms are represented pictorially in Fig. 2, and are described below.

Density.   The viscous fluid used has a density, or apparent density in between that of the minerals to be separated, such that one mineral(s) will have a net positive buoyancy and will float, whilst the other(s) will have a net negative buoyancy and will sink.   Into this classification fits heavy medium separation, one of the most important of all gravity concentration processes, as well as magnetohydrostatic separation.

110

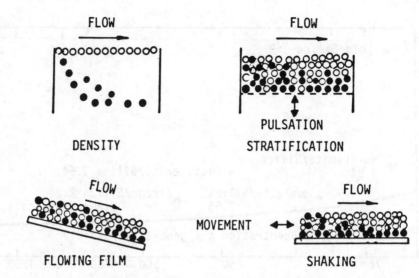

Fig. 2 The various mechanisms of Gravity Concentration

Stratification. The various mineral constituents are stratified by being subjected to an intermittent fluidization caused by the pulsation of the fluid in a vertical plane. This classification is represented by jigs of which there are a wide range and number.

Flowing Film. The various constituents are separated by their relative movement through a stream of slurry which is flowing down a plane by the action of gravity. Flowing Film Concentration is the oldest process used in gravity concentration, and remains of major importance, with such units as the sluice, or palong, pinched sluice, Reichert Cone, and, by including lateral movement caused by the addition of centrifugal force, the spiral.

Shaking. The mineral components are stratified by superimposing on the flowing film, a horizontal shear force, whether it be oscillating as in a shaking table, or orbital, as is the case on the Bartles-Mozley Separator or Crossbelt Concentrator.

The various mechanisms can be represented pictorially, as shown in Fig. 2.

UNIT PROCESSES

All concentration equipment has a primary function and a secondary function. The primary function is the efficient separation of the constituent minerals into two or more discrete phases; the secondary function is to effectively cause the discharge of these discrete phases into separate compartments (7). For maximum efficiency of operation, the forces which effect the primary function should not also result in the secondary function occurring; in such case optimum operating conditions required for the primary function, separation, may have to be compromised to ensure reasonably efficient operation of the secondary function, product discharge.

With gravity concentration, the primary function is effected by the relative movement of minerals of different density in response to gravity and one or more other forces, the latter often including the resistance to motion offered by a viscous fluid. It is a characteristic of all gravity concentration devices that in order for the particles to move relative to one another that they are, at some stage of the process, held slightly apart from each other, or 'dilated', by the superimposed force.

The previous section indicated the four mechanisms involved in gravity concentration, which effect the movement of particles relative to each other.

There is a wide range of gravity concentration equipment available to the mineral processing engineer. Some of the factors which determine the choice of the correct unit will first be discussed, followed with a more detailed look at some of the more important processes available.

Factors which are important in equipment selection include: the duty required, the size range of the particles to be separated, the throughput required, the efficiency expected and, all other things being equal, the unit cost, both capital and operating, of the equipment.

The Duty Required

Concentration of a mineral in any concentrating device is achieved by separating the valuable mineral from the waste mineral. However, with the possible exception of heavy medium separation in some specific instances, no unit process is capable of achieving a complete separation in one pass, and the unit will produce, at best, either a final concentrate or a final tailing, but rarely both.

Gravity Concentration devices are no exception to this rule and more than one stage of concentration is invariably required; these stages may be rougher-scavenger, rougher-cleaner or a combination of both: pre-concentration in this context may be equated to rougher concentration. Some gravity concentration equipment is more suited to roughing or scavenging duty, other equipment is more suited to cleaning.

Size Range

The overall size range to which gravity concentration is applied is larger than with any other process. The top size which can theoretically be treated by gravity concentration devices is the size at which different minerals are sufficiently liberated from each other for there to be a suitable gravity differential. Practically, the top size is limited by the mechanical handling ability of the device, and whilst some units are capable of handling particles as coarse as 1 metre, 500 mm is generally accepted as the top size of separation (4). The practical lower cut-off for fine gravity concentration is about 6 micrometres.

None of the equipment described later is capable of treating the whole size range of material: different types of concentrators are capable of handling but a part of the total range. Fig. 3 shows the range of operation of the more common types of equipment, split into the broad classifications stated earlier.

Not only is no equipment capable of treating the whole size range amenable to gravity concentration, few are capable of efficiently treating on one unit the whole range of which that

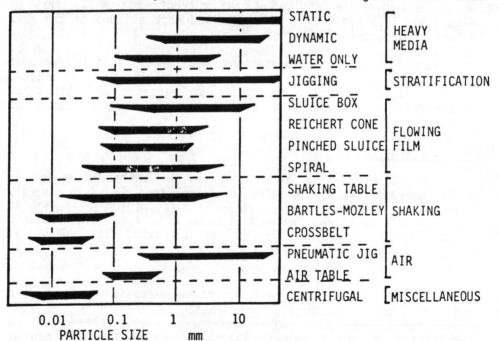

Fig. 3 Operating Range of Gravity Concentration Units

particular unit is capable of handling. Heavy medium separators which separate according to density alone are the exception. Whilst jigs, Reichert Cones and Sluices, being essentially mass classification devices, are capable of treating a fairly wide range of sizes, some loss of efficiency occurs at the top, bottom, or middle of the size range.

Other equipment such as the Table or Bartles-Mozley Separator, concentrating by reverse classification, or the spiral by hindered settling, operate significantly better on size or classified feed. This leads to two basic types of treatment circuit.

(i) Series circuits, as in sluices or dredges, treat comparatively unsized feed in successive stages each producing concentrates, middlings and tailings. Removal or rejection from circuit reduces feed to next stage. One refinement is cut-off sizing of tailings to reject barren oversize (8) and undersize fines (9).

(ii) Parallel circuits treat closely sized feed with multiple units producing concentrates, middlings and tailings in one stage. The efficiency of sizing is an important factor if finished products are required.

Generally the two types of circuits are combined. For example, a series circuit of jigs treating unsized feed may be followed by a sizing step and a parallel circuit of tables, and film sizing units.

Throughput

The capacity range of gravity concentration equipment is enormous. Whilst heavy medium separators, and jigs are manufactured in a variety of sizes, suited for a range of capacity, other items are manufactured essentially in one, or two sizes only, and high throughputs require multiple units.

The highest capacity units are static heavy medium cones or drums, and jigs, both of which can exceed 300 t h^{-1} throughput per unit, but throughput can of course be much lower with the appropriately sized unit.

Of the other units, the capacity of the Reichert Cone is of the order of 60-70 t h^{-1}; its geometry precludes its effective operation outside of a comparatively narrow range. Spirals have a capacity of 1-4 t h^{-1} per start, or with triple start units 3-12 t h^{-1}. However, the larger diameter spirals manufactured in the Soviet Union have a rated capacity as high as 7-15 t h^{-1} per start (10). Similarly, pinched sluices are rated 3-6 t h^{-1} per start.

Shaking tables, Bartles-Mozley Separators and Crossbelts have a lower capacity, the actual amount depending on individual separations and the size range to be treated. For fine material stage treatment using the Bartles-Mozley followed by either the Crossbelt or Shaking Tables will minimize equipment required and maximize throughput.

Efficiency

Of all the gravity concentration methods, heavy medium separation is, in principle, the simplest, and in practice the most efficient. The perfect gravity separation process would separate 100% of the unwanted material into one fraction and 100% of the wanted material into the other. In heavy medium separation, the criterion of separation of one particle from another is essentially related to that particle's specific gravity relative to that of the medium; in all other processes all particles are heavier than the medium and separation is determined by a complex interaction of sometimes opposing forces.

Efficiency of an individual item of equipment will depend on a variety of factors, not directly controllable by the device, such as feed flowrate, compared to the capacity of the unit, variability of flowrate or mineral make-up of the feed, size range, and solids content; the efficiency of any concentrating device will depend on the effectiveness with which the unit can handle these extraneous factors, yet maintain acceptable recovery and concentrate grade.

Heavy Medium Separation

Heavy Medium Separation (HMS) is the simplest and one of the most widely applied gravity concentration processes both in minerals treatment and in coal preparation. It is a process applied to the separation of minerals in a liquid or a fairly stable suspension of a predetermined density, chosen such that the density is higher than the lighter constituents and lower than the heavier constituents. It differs from all other gravity concentration processes, where the medium (generally water) is of a density lower than all the constituents in the ore.

There are two major areas of application for heavy medium separation. It can be used to produce a commercially saleable end product, such as in coal preparation - where it is one of the two primary means of cleaning coal from the shale - and in some industrial minerals applications. Alternatively it can be used to produce an economically acceptable waste product which does not warrant the cost of further treatment: in this case it is a preconcentration device and as such is of major importance in the preconcentration of diamonds, sulphides and metal oxides.

Unlike most other gravity concentration devices, heavy medium separation is a system rather than a unit process. To work effectively and economically, the system will consist of a series of interconnected phases:

(a) Feed Preparation
(b) Feed and medium presentation
(c) Separation of heavies and lights in a suitable vessel
(d) Product recovery
(e) Medium recovery.

These various stages are illustrated in a typical Heavy Medium Separation circuit, Fig. 4.

In operation, the feed must be screened to remove fine ore and slimes prior to it being fed, with reconstituted medium into the separatory vessel. Floats and Sinks are withdrawn separately and drained, on static or vibratory screens, of the majority of the medium which returns either direct to the system or is cleaned

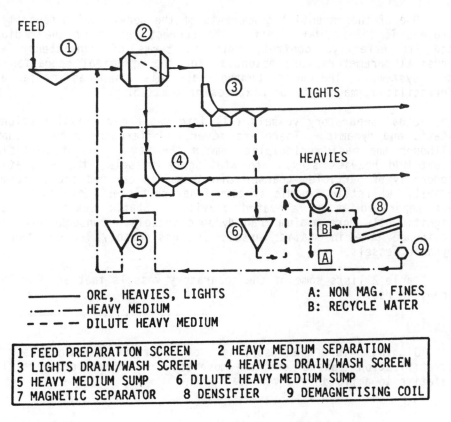

FEED

LIGHTS

HEAVIES

——————— ORE, HEAVIES, LIGHTS
.——·—— HEAVY MEDIUM
- - - - DILUTE HEAVY MEDIUM

A: NON MAG. FINES
B: RECYCLE WATER

1 FEED PREPARATION SCREEN   2 HEAVY MEDIUM SEPARATION
3 LIGHTS DRAIN/WASH SCREEN   4 HEAVIES DRAIN/WASH SCREEN
5 HEAVY MEDIUM SUMP   6 DILUTE HEAVY MEDIUM SUMP
7 MAGNETIC SEPARATOR   8 DENSIFIER   9 DEMAGNETISING COIL

Fig. 4 Typical Heavy Medium Separation Circuit

prior to return. Next, the floats and sinks are washed, on vibratory screens to remove essentially all the remaining adhering heavy medium.

The undersize products from the washing screens, consisting of medium, wash water, and fines, are too dilute and contaminated to be returned directly as medium to the separatory vessel. They are treated individually, or together, by magnetic separation, to separate the magnetic ferrosilicon, or magnetite, from the non-magnetic fines. Reclaimed, cleaned medium is thickened to the correct density by a suitable classifier, which continuously returns it to the HMS circuit. The densified medium discharge passes through a demagnetizing coil to assure a non-flocculated, uniform suspension in the separatory vessel (11).

The circuit must be regarded as a whole for both design and operational purposes. Equipment for each stage must be matched with the rest of the circuit, both in terms of capacity and performance.

One of the essential components of the heavy medium separation process is the medium itself. The correct choice of the medium, and its effective control, both in terms of consistency and physical parameters, are essential for the efficient operation of the system. The most common materials used as media are ferrosilicon, magnetite or mixtures of the two.

The actual separatory vessels fall into two broad classifications; static and dynamic. There are several differences between them, although the basic principles remain the same: light particles float and heavies sink. In stationary vessels, the separation generally of particles coarser than 3 mm, is carried out at normal gravity; whilst in dynamic systems finer particles (down to 0.5 mm) are separated at an elevated gravity. Static vessels contain significantly more medium than dynamic vessels. Consequently, the residence time in static vessels is considerably longer than in dynamic vessels.

Table 2 lists some of the separatory vessels that are, or have been, in use.

Jigging

Jigging is the sorting of different minerals by stratification, based on the movement of a bed of particles relative to a fluid in a vertical plane.

TABLE 2

Some Heavy Medium Separation Vessels

Deep Cone	Drum	S T A T I C Trough	Shallow Combination	DYNAMIC Centrifugal
Chance	Wemco	DSM Bath	Drewboy	DSM Cyclone
Barvoys	OOSK 30	McNally LoFlo	Norwalt	Swirl Cyclone
Tromp Deep	Teska	Tramp Shallow		Vorsyle
Wemco	Hardinge	Ridley Scholes		Dyna-Whirlpool
Huntindon-	Link Belt	OCC		Triflo
Heberlein	Bretby	Heyl & Patterson		Water-only
Hirst	Snail			cyclones
Washer	Neldco			Visman-Tricone

The basic construction of a jig is shown in Fig. 5. Essentially it consists of an open tank filled with a fluid, normally water, with a horizontal jig screen at the top and with a spigot in the "hutch" or bottom compartment for removal of heavies. The jig bed consists of a layer of coarse heavy particles (ragging) on the jig screen. This ragging is of a specific gravity between that of the minerals it is intended to separate.

In operation the bed is made fluid by a pulsating current of fluid to provide stratification. On the upward or pulsion stroke the bed of ragging and slurry is normally lifted as a mass, then as velocity of the upcurrent decreases it dilates, loosening the whole

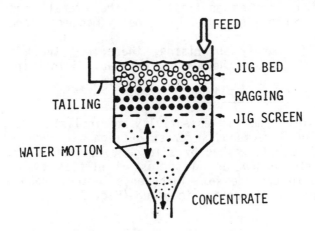

Fig. 5 Schematic Arrangement of a Jig

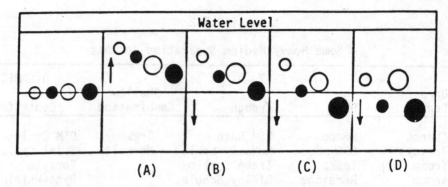

Fig. 6 Diagramatic representation of the jigging cycle, based on
four particles of different size and density
(A) pulsion (B) differential acceleration
(C) add hindered settilng (D) add interstitial trickling

mass. On the downward or suction stroke it closes slowly again.
The ideal is to control this dilation of the bed of material so
that the heavier and smaller particles penetrate the interstices of
the bed, whilst the larger heavies fall under a condition akin to
"hindered" settling. Fig. 6 shows diagramatically the effect of
the various mechanisms of jigging on four particles, of different
size and density.

Whilst jigs were once applied to the concentration of a wide
range of minerals, the major modern applications of jigs are
essentially (a) coal washing, (b) primary treatment of alluvials,
(c) coarse lode tin ore recovery, (d) free metal recovery in gold
and other mills, (e) cleaning of sand. Nevertheless even with
these more limited applications, the overall tonnage treated by
jigs is probably second only to heavy medium separation.

It is hardly surprising, therefore, that there are a wide
variety of jigs on the market, and Table 3 indicates some of these.

The Sluice Box, or Palong

A sluice box is essentially an inclined trough, or launder,
usually of a rectangular cross section, through which a long size
range of feed is washed by a large volume of water. In the bottom
of the sluice box are a series of riffles of various types, so
arranged to provide some turbulence between each riffle, allowing
concentration of heavies to occur (Fig. 7).

TABLE 3

Types of Jigs, With Typical Examples

| Type | Method of Pulsation | Mineral Jigs | | Coal Jigs | |
		Over the Screen	Through the Screen	Over the Screen	Through the Screen
Moveable Screen		Halykyn James	HANCOCK	Wilmot Pan	
Fixed Screen Mechanical	Plunger	HARZ Woodbury McLanahan Stone	Cooley Collom May	ORC Elmore Reading	Faust
	Diaphragm	BENDELARI RUOSS	DENVER WEMCO/REMER YUBA RICHARD Pan Am.Placer Panam-Kraut IHC	JEFFEY	
Pulsator	Air	OPM Series		BAUM BATAC Tacub	FELDSPAR Cortex
	Water Vane	Richard	Pam American Neil		Vissac

Whilst the majority of sluice box operators use manual "clean-up" - or removal of concentrates, some attempt to mechanise this time consuming phase has been made in the U.S.S.R. (12).

SLUICE BOX

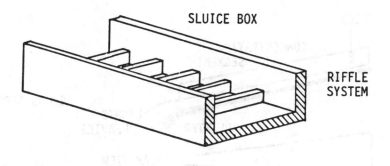

RIFFLE SYSTEM

Fig. 7   Cross Section of a Simple Sluice-Box

A sluice box is a complete placer pre-concentrating plant; it is a combination of a feed hopper, conveyor system and gravel washing and heavy mineral saving apparatus (13).

Although some manufacturers do offer sluice boxes, the majority are shop fabricated for specific operations; the design and construction therefore depends on personal preference as well as the duty to which the sluice is to be put.

The Pinched Sluice and Reichert Cone

The pinched sluice differs from the riffled sluice both in the smooth bottom of the trough and in the method of concentrate removal. In the riffled sluices heavies settle into the gaps between the riffles set at right angles to the flow and are removed intermittently, whereas in a pinched sluice the flow regime is such that heavies are removed continuously.

The general principle of opertion of a typical sluice type of separator is shown in Fig. 8. It normally consists of an inclined channel which decreases in width in the direction of the flow. Pulp, at a high solids density enters the channel in a relatively thin stream at the upper, or wider, end. As this stream flows down the sluice a normal flowing film velocity gradient is set up and the finer and heavier particles concentrate in the lower levels of the stream, by a combination of hindered settling and interstitial trickling. In the normal form of sluice, where the channel width decreases in the direction of flow, the depth of the stream increases roughly in proportion to the rate of pinch, and this increases bed thickness and naturally enhances the separation of heavy and light materials.

At the end of the channel the slower moving lower level of the stream, suitably enriched in heavy minerals content, is peeled from

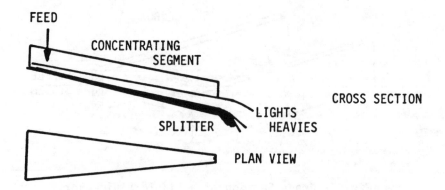

Fig. 8 Principle of Operation of a Pinched Sluice

the stream by a suitable cutter. The faster moving upper layers, depleted in heavy minerals, pass over the cutter to report as a separate product.

Early development of the pinched sluice was in the United States; however it was in the Australian Mineral Sand industry that they truly became popular. Table 4 indicates some of the range of Pinched Sluice type units that are, or have been available.

Reichert cones (Fig. 9) are a special type of pinched sluice (14, 15). The inverted cone concentrating surface may be viewed as a number of pinched sluices placed adjacent to one another to form a circle. The one major difference is that in the cone there are no sidewalls to the sluice to interfere with stratification. In essence this is the only difference between a number of pinched sluices and a cone. A cone concentrator always consists of a number of stages in series and in parallel and can thus achieve multiple stages of separation in one unit. This is important as neither cone nor sluice is particulary efficient in a single stage.

The advantages of the Reichert cone are its high capacity, low capital cost per unit throughput, light weight, compactness and metallurgical performance. Cones have two disadvantages, one of which is shared with sluices. This latter is the necessity of feeding at 60% solids as both units are sensitive to variations in pulp density. The other disadvantage is that cones are unsuitable for operations with a new feed rate of less than about 50 t h^{-1}. Cones have an upper size limitation of about 3 mm on feed, while the lower size limit for particle recovery is about 40 micrometres depending upon particle specific gravity. Just what recovery is obtained for 40 micrometre particles depends very much on the mineral, but is generally greater than 50%.

Table 4

Typical Pinched Sluice Type Units

U.S.A.	AUSTRALIA	SOUTH AFRICA	U.S.S.R.
Cannon	York	Rand Leases	Giridmet
Carpco	Belmont	Plane Table	
Lamflo	Cudgen		
Hobart	RZ		
	Diltray		
	Xatal		
	Wright Impact Tray		
	Reichert Cone		

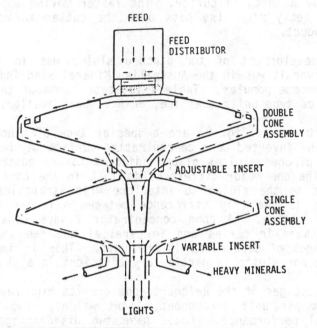

FEED

FEED DISTRIBUTOR

DOUBLE CONE ASSEMBLY

ADJUSTABLE INSERT

SINGLE CONE ASSEMBLY

VARIABLE INSERT

HEAVY MINERALS

LIGHTS

Fig. 9   Internal Flow Diagram of one "DS" Stage
of a Reichert Cone Concentrator

Spiral Concentration

The spiral concentrator is one of the relatively modern, high capacity low cost units developed for the inexpensive preconcentration of low value ores. The spiral is a unique device in its approach to separation and recovery. Although modern units are available in a variety of designs, the basic principles of spirals have hardly changed since their introduction in the 1940's.

The development of the spiral concentrator typifies the ingenuity and inventiveness of practical mineral processing engineers. In 1941, Humphreys Gold Corporation Vice President, I.B. Humphreys, investigated the possibility of development of an inexpensive method of gold recovery from their operations. The initial experiments utilized old automobile tyres and later, sheets of lead to produce a spiral configuration. These experiments with flexible materials allowed Humphreys to test various diameters, pitches and cross sections (16).

Until very recently, all spirals were similar; each consisted of a multi-turn helical sluice of modified semi-circular cross-section, for pulp flow, with a wash water channel and a series of concentrate off-take ports at regular intervals down the spiral.

Recently, however, there has been considerable development in spiral technology; the main areas of development are modified helix cross sections and diameter, the development of (wash) waterless spirals, and the introduction of spirals with only one take off point, at the bottom of the spiral.

Nevertheless, all spirals are similar in principle. Feed pulp of between 20-25% solids by weight and in size range 3 mm down to 50 micrometres is introduced at the top of the spiral and, as it flows spirally downwards, the coarsest and heaviest particles concentrate in a band along the inner side of the stream (Fig. 10).

Concentration in the spiral may be considered as involving three inter-dependent and co-existent concepts of mineral separation: (i) free settling, (ii) fluid film, and (iii) stream cross-sectional rotation. Many gravity concentrating devices utilize free settling and flowing film principles but the addition of cross-sectional rotation makes the spiral unique as a self compensating device.

There has been a considerable development in spiral technology since the original Humphreys design. Manufacturers now include, apart from Humphreys Engineering; GEC, Budin, Sala (Europe); Reichert, Vickers, Wright, Wyong (Australia); Spargo (South Africa); the SVM, PLSM and ShBM spirals (U.S.S.R.).

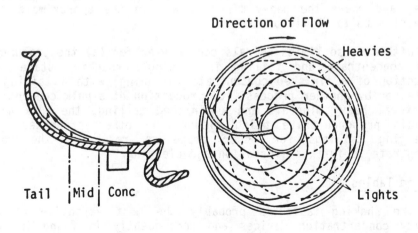

Fig. 10 Cross Section of a Spiral, showing particle movement

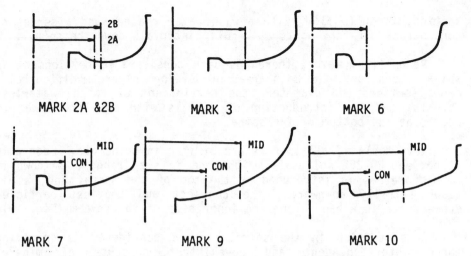

Fig. 11   Reichert Spiral Cross Section

Typifying the development of the helix cross sectional shape is the range of models offered by Reichert (17), as shown in Fig. 11.

Outside of the Soviet Union the spiral is typically 0.6-0.7 m in diameter; however Soviet engineers have developed a range of spirals ranging from 1-2 mm in diameter (10, 18) including one, novel, 9 start spiral.

Spirals can be employed for the separation of any two minerals from each other, as long as the minerals are essentially liberated, there is a reasonable difference in specific gravity or particle shape, and where the heavy minerals are in the approximate size range of 3 mm to 50 micrometres.

Within these limits spirals can be used for (a) the production of a concentrate and tailing in a single roughing stage, (b) production of a finished concentrate (cleaning) with a tailing to be treated by another process, (c) production of a bulk concentrate of several heavy minerals and a finished tailing, the bulk heavy minerals preconcentrate being cleaned by other processes, (d) scavenging tailings for other processes returning the rough concentrate to further liberation and cleaning.

Shaking Tables

The shaking table is probably the most versatile of all gravity concentration devices and consequently is found in most types of gravity mill. Almost certainly any well equipped mill laboratory will boast a small laboratory table for that occasional test.

The shaking table consists of a slightly inclined deck on to which feed, at about 25% solids by weight, is introduced. The table is vibrated longitudinally, using a slow forward stroke, and a rapid return, which cause the mineral particles to "crawl" along the deck parallel to the direction of motion. The minerals are subjected to two forces - that due to the table motion and that, at right angles to it, due to a flowing film of water. The net effect is that the particles move diagonally across the deck from the feed end and, since the effect of the flowing film depends upon the size and density of the particles, they will fan out on the tables, as shown in Fig. 12.

Good classification will invariably optimize operation of tables, especially where high enrichment from different ores with a fair proportion of middlings is required. Poor classification on the other hand may even be detrimental.

Since the introduction of the Wilfley Table in 1896, a wide variety of shaking tables have been marketed. Currently, however, there are only a few major manufacturers of shaking tables, although several other less known types are operating in some locations.

The major differences between tables are the shape and suspension of the deck, and in the type of mechanism which imparts the assymetric reciprocating motion to the deck.

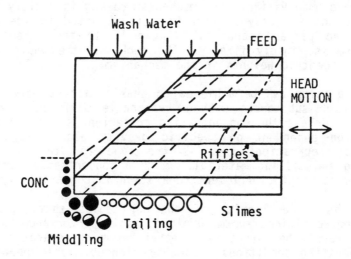

Fig. 12 Particle Movement on a Shaking Table

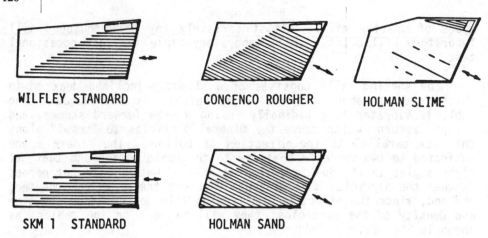

WILFLEY STANDARD   CONCENCO ROUGHER   HOLMAN SLIME

SKM 1 STANDARD   HOLMAN SAND

Fig. 13 Riffle Patterns on Various Tables

There are two basic shapes of table deck: rectangular and diagonal. Fig. 13 shows two rectangular type table decks (Wilfley and SKM) and three diagonal (Concenco and Holman). It should be noted that the "longitudinal" motion is in each case parallel to the riffle pattern.

The diagonal deck generally has a higher unit capacity, produces a higher grade of concentrate over a wider band, and reduces the quantity of middlings. A diagonal deck will also recover finer size particles.

There are two basic types of mechanism, or head motion. In the more common variant, the mechanism casing is rigidly fixed to a subframe and the assymetric motion is imparted to the deck by a toggle and pitman mechanical linkage. In the other type, the mechanism is rigidly fixed to the deck and the whole shaken, as one, by eccentric weights in the mechanism.

Separation can be achieved on shaking tables between any two minerals or substances with a reasonable difference in specific gravity. Unless the concentration criterion is greater than 1.25, separation dependent on specific gravity alone will result in imperfect separation; such is the case for the separation of middling and tailing material. With a concentration criterion of 2.5 or more, rapid, efficient separation is comparatively simple.

It is a peculiarity of shaking tables that, when correct operating conditions are maintained, table performance is excellent and can rarely be surpassed by other types of separators; however, when operating conditions become unbalanced, table performance is

TABLE 5

Parameters Affecting Operation of Shaking Tables

Machine	Feed
Stroke: length and frequency Inclination: side and longitudinal Riffle Pattern Deck Surface	Solids Characteristics - Size Range - Capacity - Cut Points Pulp Characteristics - Flowrate - Density - Wash water

likely to be as far on the bad side as it was on the good side under favourable conditions (19). With anything but the simplest of operation, therefore, table performance can be either extremely good or extremely bad; which, depends on adherance to correct operating conditions.

Parameters affecting table operation are shown in Table 5.

The major modern use of the shaking table is the coal cleaning industry of North America with the order of 50 Mt being treated annually (7). Elsewhere, there are large numbers of tables treating the metal oxides, chromite, mineral sands and gold.

Slime Plants

The word "slimes" in this context should be taken to mean material that is finer than 50 micrometres. There is some confusion in the mineral processing industry as to the definition of the word slimes (20). In many instances, the first stage in the treatment of "slimes" includes thorough desliming!

The inherent difficulties in recoverying very fine particles by gravity has led to the development, and in many cases, obsolescence of a wider and more diverse range of equipment than in any other particle size range. Table 6 highlights this diversity of equipment. The important modern devices are highlighted.

The Endless Belt Concentrator and Johnson Barrel are commonly used in South Africa for the recovery of fine gold in milling circuits; they are, however, not used elsewhere. Likewise, the Rocking Shaking Vanner and the Yunnan Centrifugal Separators are used only in China, for the treatment of tin, tungsten and iron ores. Of all the modern "slimes" separators probably the most

TABLE 6

Range of "Slimes" Gravity Concentration Equipment

STATIONARY DECK EQUIPMENT	STIRRED BED DEVICES	CENTRIFUGAL DEVICES
-Buddle	Discontinuous Shear	-Ferrara's Tube
-Round Table	-Vanners	-DTsS Centrifugal
-Round Frame	-Shaking Table*	-Yunnan Separators*
-Strake	-Kieve	-Knelson Hydrostatic
-Corduroy Table	-Rocking Shaking Vanner*	-TsBS Centrifugal
-McKelvey Concentrator	-GEC Duplex Concentrator	
-Denver Buckman Tilting Frame		
	Unidirectional Shear	
	-Endless Belt Concentrator*	
	-Johnson Barrel*	
	-Hodgson Separator	
	-Rotating Cone Separator	
	Orbital Shear	
	-Shaken Helicoid	
	-Bartles-Mozley Separator*	
	-Bartles Crossbelt Concentrator*	

widely used are the Bartles-Mozley Separator (21) and the Bartles Crossbelt Concentrator (22). Both units superimpose, on a slightly inclined surface an orbital shear motion, and are capable of recoveries as fine as 6 micrometres. The Bartles-Mozley Separator (Fig. 14) is a semi batch device, discharging concentrates discontinuously, whilst the Bartles Crossbelt concentrator Fig. 15) is a continuous device, using a slowly moving belt to remove concentrates from the pulp stream.

The newest devices, the GEC Duplex Concentrator and the Knelson Hydrostatic separator show promise for the future, but as yet have only had comparatively limited application.

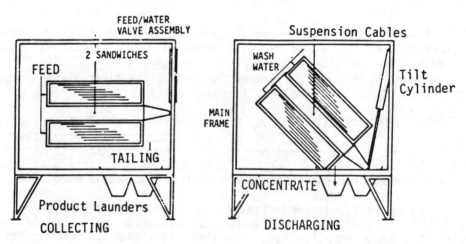

Fig. 14 Operating Modes of the Bartles-Mozley Separator

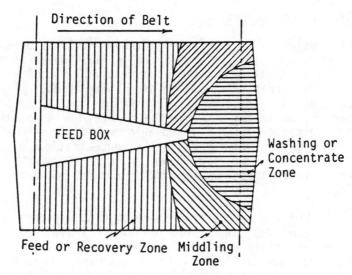

Fig. 15   Schematic Plan of Bartles Crossbelt Concentrator

Treatment of slimes usually requires stage treatment, as enrichment in any only stage is never particularly high. Fig. 16, a schematic flowsheet of the gravity circuit at the Solnenchnyy concentrator in the Soviet Union (23) illustrates this.

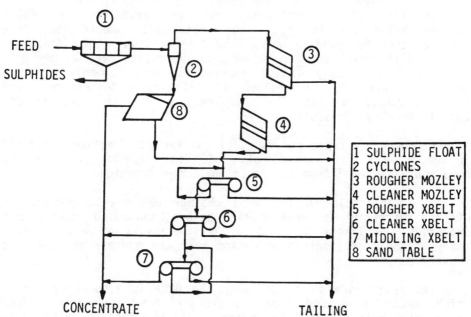

1	SULPHIDE FLOAT
2	CYCLONES
3	ROUGHER MOZLEY
4	CLEANER MOZLEY
5	ROUGHER XBELT
6	CLEANER XBELT
7	MIDDLING XBELT
8	SAND TABLE

Fig. 16   Complex Gravity Circuit for the Concentration of Tin
at the Solnenchnyy Concentrator USSR

Dry Gravity Concentration

The majority of dry gravity concentration processes that have been developed, or are commercially available, are similar in principle to their wet gravity concentration counterparts. These include the areas of heavy medium separation, jigging, pinched sluices, and tabling. Whilst the mechanisms of separation of the wet and dry units may not necessarily be the same, the wet units, already described make a reasonable point of reference for their dry counterpart. Other dry gravity concentration devices have been developed which do not have a direct counterpart in wet concentration. These include the Fluidized Bed Separator, and the Zig-Zag.

All the dry concentrating equipment, except the dry jig, utilize a constant controlled upflow of fluidizing air. The dry jig, however utilizes a pulsating air flow, using the pulsation to effect the stratification in a manner comparable with its counterpart in wet concentration.

TESTING FOR GRAVITY CONCENTRATION

Whilst this has been discussed in more detail elsewhere (eg. 3, 24) a brief introduction is warranted.

The first step at the laboratory level, assuming that a fully representative sample has been obtained, should be size fractionation followed by heavy liquid separation of the fraction. A prior size reduction step to enhance liberation may, or may not, be necessary. Heavy liquid separations are commonly made in beakers, separatory funnels or in batch-type centrifuges. A complete sequential sink-float analysis of the proposed process feed is usually advisable.

A thorough microscopic study of the various fractions produced in the heavy liquid work is also of vital importance. Probably more can be learned from this than any other technique.

It is possible to determine whether wet gravity separation will be successful in beneficiating an ore, and which gravity unit processes are likely to succeed, on the basis of heavy liquid analysis and microscopic examination over the appropriate specific gravity range.

The next stage is to scale-up the data to laboratory scale, which means the production of larger feed samples. Feed preparation must also be considered. This may include grinding to liberation size, desliming, screening to remove unwanted oversize, hydraulic classification or a combination of the above.

Most manufacturers provide small scale versions of their production equipment although results, especially in the fine particle range must be treated with care. Some equipment (especially spirals) should properly be regarded as pilot plant units.

The question of wet gravity separation pilot plant testwork is always controversial, particularly if a continuous pilot plant is envisaged.

PLANT OPERATIONS

With the wide range of minerals treated, the equipment employed and the degree of sophistication acceptable, there cannot be any "typical" plant operations. Two plants, one treating iron ore and the other tin are included here as examples of two very different types of operation.

Mount Newman Concentrator, Australia

As with other West Australian iron ore producers, the Mount Newman mine commenced operation on its large high grade ore zones, using simple crushing and screening. However, it has more recently installed a wet concentrator to treat the lower grade zones of the orebody (25).

Feedrate to the concentrator is approximately 1,160 t h^{-1} which is just over one tenth of the total output of the mine.

A considerable amount of testwork was carried out, on various competing processes prior to making the final decision to proceed with the concentrator. All ore fractions exhibited minimal near density material between the high density hematite and the low density shales. Furthermore, the absence of any composite hematite/shale particles as coarse as 100 mm indicated that sharp separations could be achieved with static heavy medium separation. The similar washability characteristics of the ore throughout the size range of -100 +6 mm precluded the necessity of providing separate facilities for the treatment of this size range.

A schematic flowsheet of the concentrator is given in Fig. 17. Run of mine low grade ore is crushed to 100 mm in a two stage crushing circuit with a rated capacity of 2000 t h^{-1} discharging to a bank of dry screens, screening at 6 mm, screen products passing to small, 1000 t surge bins. Lump ore, -100 +6 mm in size is treated in two Wemco Heavy Medium drums in parallel producing lump concentrate and tailings, which after rinse:drain screening pass to the appropriate stockpiles. Screen undersize is further wet screened, at 1 mm, oversize passing to a surge bunker

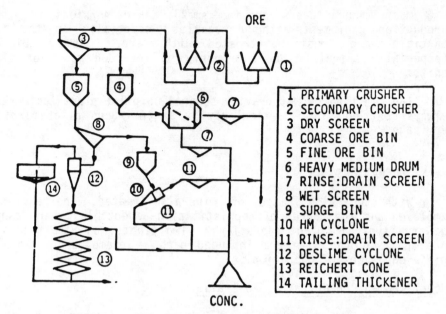

Fig. 17   Schematic Flowsheet of Mount Newman
          Concentrator, Australia

and thence to three DSM heavy medium cyclones in parallel.
Products pass to rinse:drain screens and thence to product
stockpiles. The -1 mm fraction is deslimed, at 63 micrometres in
cyclones, cyclone overflow passing to a thickener for water
recovery. Cyclone underflow, at up to 280 t h^{-1} passes to a
bank of Reichert Cones, producing a concentrate and tailing.

All concentrates are stored in one stockpile, for subsequent
final crushing and sizing into lump and fine ore at dockside.

Geevor Tin Mine: Concentration Circuit

Geevor's concentration circuit is an excellent example of an
efficient, all gravity concentrator, combining effective
classification, coarse and fine tabling, and an efficient slime
recovery circuit. It is shown schematically in Fig. 18 (most
concentrate and tailing flows are omitted for clarity). The
circuit exemplifies the recovery of coarse free cassiterite
followed by careful regrinding and retreatment of middlings (26).

Crushed ore is fed from the 1000 t fine ore bin at 18 t h^{-1}
to two Allis Chalmers low-head screens in parallel, fitted with
polyurethane screens and sizing at 0.85 mm. Screen oversize is
ground in a 2.2 m dia. x 2.25 m rubber lined grate discharge ball

mill, the product from which recycles to the screens. Screen undersize is fed to the new table plant classification system, which consists of a full flare spiral classifier and Stokes 8 spigot hydrosizer in series, with the combined overflow being further classified in 250 mm and 150 mm cyclones in series. All spigot products are treated on a total of 31 Wilfley Holman tables.

Coarse table middlings, (from the Stokes hydrosizer) are reground in open circuit, reclassified in a spiral classifier and Stokes 10 spigot hydrosizer and treated on a bank of 10 Holman coarse and fine sand tables. Middlings from these tables are again reground in open circuit and then join the middlings from the fine and slime tables treating the cyclone underflows. The combined flow is again classified, in a Stokes 7 spigot hydrosizer prior to tabling on a bank of 7 Holman coarse and fine sand tables. Middlings from these tables are ground in open circuit prior to recycling to the 7 spigot hydrosizer. Each of the above middling regrind circuits consists of a dewatering spiral or rake classifier and a conical ball mill.

Overflows from the classifiers are thickened in a 5.5 m dia. thickener the sands from which are stage cycloned in 250 mm and 125 mm dia. cyclones, cyclone underflows being treated on a bank of Holman fine sand tables and slime tables.

Fines from the crushing plant (cyclone overflows) are thickened in a 6.4 m dia. thickener the underflow of which is classified in a Stokes 3 spigot hydrosizer with the products being treated on a bank of Holman fine sand tables. Middlings from the above two banks of tables are reprocessed in the secondary regrind circuits.

The 5.5 m and 6.4 m dia. thickener overflow, along with the 150 mm cyclone overflow passes to a 10.7 dia. thickener, the underflow of which is classified in a bank of 100 m cyclones, the spigot product being treated on a bank of 5 Holman fine sand tables; 10.7 m dia. thickener and 100 mm cyclone overflows are again thickened in a 21.5 m dia. thickener, the overflow of which is recycled as process water. The underflow is deslimed in 100 mm cyclones prior to treatment on a bank of Holman slime tables.

Tailings from the tables are processed on banks of Bartles-Mozley Separators and shop fabricated Mozley Frames - prototype units for the commerical devices - in series. Rough concentrates are upgraded by Bartles Crossbelt concentrators and Holman slime tables in parallel. It is noteable that Geevor's flowsheet contains the prototype of both the Bartles-Mozley separator and the Crossbelt, as well as the Jones separator and Mintek analyser; a fine record of innovative processing in this old established mill.

134

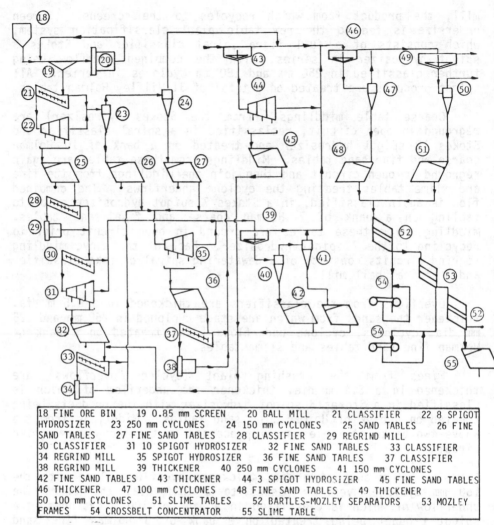

18 FINE ORE BIN	19 0.85 mm SCREEN	20 BALL MILL	21 CLASSIFIER	22 8 SPIGOT
HYDROSIZER	23 250 mm CYCLONES	24 150 mm CYCLONES	25 SAND TABLES	26 FINE
SAND TABLES	27 FINE SAND TABLES	28 CLASSIFIER	29 REGRIND MILL
30 CLASSIFIER	31 10 SPIGOT HYDROSIZER	32 FINE SAND TABLES	33 CLASSIFIER
34 REGRIND MILL	35 SPIGOT HYDROSIZER	36 FINE SAND TABLES	37 CLASSIFIER
38 REGRIND MILL	39 THICKENER	40 250 mm CYCLONES	41 150 mm CYCLONES
42 FINE SAND TABLES	43 THICKENER	44 3 SPIGOT HYDROSIZER	45 FINE SAND TABLES
46 THICKENER	47 100 mm CYCLONES	48 FINE SAND TABLES	49 THICKENER
50 100 mm CYCLONES	51 SLIME TABLES	52 BARTLES-MOZLEY SEPARATORS	53 MOZLEY
FRAMES	54 CROSSBELT CONCENTRATOR	55 SLIME TABLE

Fig. 18 The Geevor Mine Concentration Circuits

ACKNOWLEDGEMENTS

Material for this paper is taken from the new textbook "Gravity Concentration Technology" written by this author (with the assistance of C. Mills) and which is being published by Elsevier Science Publishers B.V. of Amsterdam, who retain the copyright. Their permission to use the material for this paper is duly acknowledged.

Permission to present this paper, given by the Directors of Tantalum Mining Corporation of Canada is also gratefully acknowledged.

REFERENCES

(1) Pliny, C.P.S. (circa 70). Natural History Book 33, 21.

(2) Agricola, G. (1556). De Re Metallica Trans Hoover, H.C. and Hoover, L.H., Dover Publications, N.Y. 1950 Book XIII.

(3) Mills, C. (1978). Process Design, Scale-Up and Plant Design for Gravity Concentration. In Mular, A.L. and Bhappu, R.B. (Eds.). Mineral Processing Plant Design AIME New York, Chapter 18 404-426.

(4) Burt, R.O. and Mills, C. (1982). Gravity Concentration - a process at the Crossroads. Paper presented at 111th Meeting of the Am. Inst. Min. Eng. Dallas, February 12 pp.

(5) Nio, T.H. (1978). Mineral Dressing by IHC Jigs. Paper presented at Gravity Separation Short Course. Reno, Nv. 46 pp.

(6) Kelly, E.G. and Spottiswood, D.J. (1982). Introduction to Mineral Processing. 491 pp.

(7) Aplan, F.F. (1975). The State of the Art and the Future of Gravity Concentration. In Somasundaran, P. and Fuerstenau, D.W. (Eds.) Research Needs in Mineral Processing. Harriman, N.Y.

(8) Harris, J.H. (1959). Serial Gravity Concentration - a new tool in mineral processing. Trans. Inst. Min. & Metall. 69 85-94.

(9) Chaston, I.R.M. (1962). Gravity Concentration of Fine Cassiterite. (5th) Inter. Miner. Proc. Cong. London I.M.M. 593-609.

(10) Bogdanov, O.S. (1983). Textbook of Ore Dressing - Basic Processes. Nedra Press - Moscow. 381 pp. (In Russian).

(11) Gochin, R.J. and Smith, M.R. (1983). Dense Medium Separation - an Introduction to the Theory and Practice. Min. Mag. Dec. 453-460.

(12) Cook, D.J. (1954). Gold Recovery in a Sluice Box. Eng. of Mines Thesis Univ. of Alaska (unpublished).

(13) Zamyatin, O.V., Lopatin, A.G., Sannikova, N.P. and Chugunov, A.D. (1975) The Concentration of Aureferous sands and Conglomerates Moscow, Nedra Press, 260 pp. (In Russian).

136

(14)   Ferree, T.J. (1972).   Introduction to the Reichert Cone.
       Paper presented at AIME-MBB annual meeting, Colorado
       Springs, Colo., 15 pp.

(15)   Ferree, T.J. and Robinson, C.N. (1982).   The Reichert Cone
       Concentrator a new approach to fine gold recovery in
       placer gold mining.   Paper presented to 2nd Annual
       Seminar on Placer Gold Mining - Vancouver, Feb. 11 pp.

(16)   Humphreys, I.B. and Hubbard, J.S. (1945).   Where Spirals
       replaced Tables, flotation cells.   Eng. & Min. J. 146 (3)
       March pp. 82-4.

(17)   Holland-Batt, A.B., Balderson, G.F. and Cross, M.S. (1982).
       The Application and Design of Wet Gravity Circuits in the
       South African Minerals Industry.   J.S. Afr. Inst. Min. &
       Metall.   March 53-70.

(18)   Burt, R.O. and Yashin, A.V. (1984)   Spiral Concentration -
       current Trends in Design and Operation.   To be presented
       to Mineral Processing and Extractive Metallurgy Kunming,
       China Oct.

(19)   Deurbrouck, A.W. and Palowitch, E.R. (1968).   Wet
       Concentrating Tables.   In Leonard, J.W. and Mitchell,
       D.R. (Eds.)   Coal Preparation 3rd Edn. Am. Inst. Min.
       Eng. 10-32 to 10-58.

(20)   Burt, R.O. (1968). Discussion of Collins, D.N., Kurup, J.L.,
       Davey, M.H., and Arthur, C., Flotation of Cassiterite:
       Development of a Flotation Process.   Trans. I.M.M. Inst.
       Min. & Metall. 77 C180.

(21)   Burt, R.O. and Ottley, D.J. (1973).   Developments in Fine
       Gravity Concentration using the Bartles-Mozley Table.
       Proc. Ann. Meet. Canadian Mineral Processors 5th, Ottawa
       29 pp.

(22)   Burt, R.O. (1975).   Development of the Bartles Crossbelt
       Concentrator for the Gravity Concentration of Fines.
       Inter. J. of Miner. Proc. 2 219-234.

(23)   Reid, T.G. (1983) Personal Communication.

(24)   Burt, R.O. (1984) Gravity Concentration - from Bench Scale
       to Plant.   Proc. 16th Ann. Meet. Canadian Mineral
       Processors, Ottawa 483-506

(25) Uys, J. and Bradford, W.H. (1981) The Beneficiation of Iron
          Ore by Heavy Medium Separation.   2nd Inter. Iron Ore
          Symposium Franfurt, Metal Bull. 21 pp.

(26) Lawlor, F.J., (1983) Private Communication.

PROCESSING OF MINERAL ORES BY MODERN MAGNETIC SEPARATION TECHNIQUES

M.R. Parker

Separation Science Group, University of Salford, Salford
M5 4WT, U.K.

Abstract

A review is given of the current state of development of
magnetic separation technology in relation to the field of mineral
processing. The subject matter is classified in terms of separation
devices based upon particle entrapment and upon particle deflection.
Of these particular mention is made of the Jones and of the Kolm-
Marston separators (particle entrapment) and also of open gradient
magnetic separation (particle deflection).

An assesment is made of the influence of cryogenic magnets on
the development of this technology and some reference is also made
to unit process costs for the various devices.

Introduction

The origins of magnetic separation in relation to the civilised
world are a matter of some speculation. Fragmented sources of information indicate some activity as far back as the ancient Chinese and
Middle Eastern civilizations. A more restricted historical approach
is probably appropriate to the material of this chapter since it
indicates that the origins of magnetic separation are closely linked
with the industrial revolution of the late 18th century in Europe
and the corresponding mid-19th century events in North America.
Since its beginnings magnetic separation has been intimately linked
with the processing of mineral ores and, in particular, with iron
ore. Indeed, the first recorded patent in magnetic separation, by
William Fullarton in England in 1792, relates to the upgrading of
iron ore. The first corresponding patent filed in North America, by

Ransom Cook in 1849 also concerns iron ore. Most patented and published material in magnetic separation from these early days until the beginning of the 20th century is concerned with the concentration of iron ores. The first half of the 20th century saw the arrival of a plethora of commercial magnetic separation devices for the processing of mineral ores. Most of these, such as the wet and dry drum separator, the wet and dry belt separator, the induced roll separator are still in successful commercial operation at the present day [1,2,3,4].

The diverse nature of magnetic separation techniques employed in commercial devices of the type mentioned above has led to various schemes of classification. Taggart [2] and DeVaney [5] and Parker [4] have discussed this classification problem in terms of parameters such as field gradient, separation medium (wet and dry), mode of product disposal (gravity, belt, spray, scraper,feed presentation (pulley, belt, shaking tray) and so on.

This particular review is concerned with modern magnetic separation techniques which are either in use in mineral separation or are in advanced stages of development. In the post 1970 period, magnetic separation has witnessed changes so rapid and so profound that no single text has been written which deals with them effectively and comprehensively. For the major part, modern magnetic separation systems may be codified in the following simple manner. The first of two major classes of separator may be regarded as particle entrapment devices. In these, feed material comes in direct contact with a magnetic capture surface. The more magnetic component of the feed is held at the surface, primarily, by magnetic traction forces and the less magnetic (or non-magnetic) component(s) is removed from the neighbourhood of the capture surface by one or more of a variety of competing forces. The latter include hydrodynamic drag, inertia and gravity. At some convenient point in time the process is halted and the magnetic component is removed from the capture surface to a point of collection. In the other major class of separator, the separation principle is that of selective particle deflection in a magnetic field. Here, there is no magnetic capture surface. Instead, such devices rely on the more magnetic particles following different flight trajectories from those of less magnetic particles in some non-uniform magnetic field system. In contrast to some of the entrapment devices, these deflection devices are wholly continuous, the various components of the feed stream being finally divided at a geometrical baffle.

In the following, the fundamental design principles of these two classes of device are discussed in turn and a review is given of recently published material on the application of these devices, to mineral processing.

Particle Entrapment Devices

This type of device generates, in a separation volume or zone, strong magnetic traction forces by virtue of a combination of high magnetic fields and field gradients. Since about 1971, the popular expression 'high gradient magnetic separation', often abbreviated to 'HGMS' has been coined as a description of this type of device but it may be applied equally well to somewhat older devices. What all these devices do have in common is, within the separation zone, a ferromagnetic matrix at the surfaces of which are indeed strong induced magnetic dipolar forces. It can be shown that the magnetic traction force experienced by paramagnetic particles of volume, υ_p and of volume susceptibility, χ_p, in a non-uniform magnetic field is (approximately)

$$F_m = \chi \upsilon_p (\mu_o H \nabla H) \tag{1}$$

where ∇H is the spatial gradient of the total magnetic field H (H is measured in Am^{-1}) at the location of the particle, and $\chi = \chi_p - \chi_m$ where χ_m is the susceptibility of the surrounding medium. Here, χ_p and υ_p are exclusively, properties of the particle while $(\mu_o H \nabla H)$ - which has the dimensions of force density - depend only on the properties of the magnetic field system. Modern particle entrapment devices have been designed to process small particles of weakly magnetic ores and, therefore, of necessity, (eqn.1) required to generate high force densities in the separation zone.

In the field of mineral processing, there are two quite distinct and highly successful modern magnetic separators of this type. The first of these, patented in 1955 [6] and first described in detailed scientific terms in 1960 [7] is the Jones Separator . This device, together with near equivalents, is manufactured in several countries, though the world market is dominated by the West German organisation, Humboldt Wedag. This is undoubtedly, the world's most commercially successful modern magnetic separator. Over 90 of the enormous DP317 Jones devices (see below) have been sold world-wide.

In its simplest laboratory form the Jones separator comprises an array of grooved ferromagnetic plates in the field gap of a classical dipolar C-shaped magnet. In the grooved plate matrix of Fig.1, the plates are loaded in boxes with the axis of the groove vertical. Wet feed material (pulp) enters the matrix vertically downwards under gravity in narrow gaps between the plates. When the magnetic field is removed scouring water can also be fed vertically downwards to remove magnetic material completely from the plates.

Commercially, large-scale forms of the Jones separator usually have, depending on the size of the installation, either one or two annular assemblies of grooved plate matrices (7) mounted on a

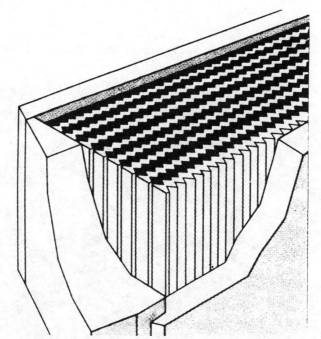

Fig.1 Grooved plate matrix of Jones Separator (Courtesy of
KHD Humboldt Wedag AG)

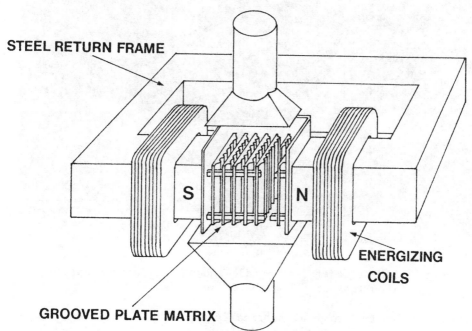

Fig.2 Schematic of laboratory Jones Separator

142

circular horizontal rotor (6) with an axial vertical drift shaft (4)
supported by heavy-duty roller bearings.   These annular matrices

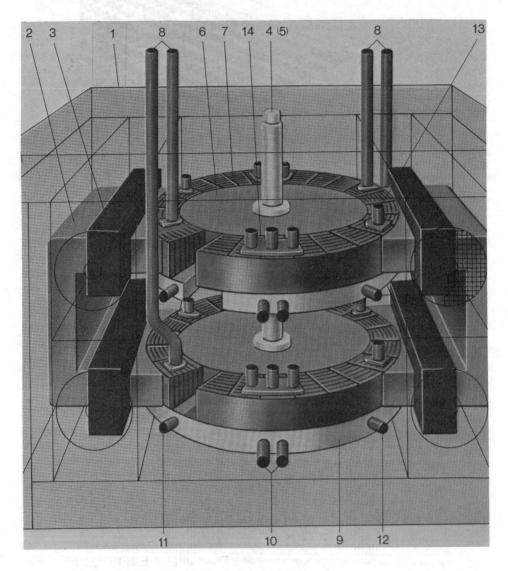

Fig.3   Schematic of large-scale industrial Jones Separator
(Courtesy of KHD Humboldt Wedag AG)

rotate between two large diametrically opposing air-cooled C-shaped
electromagnets (2,3) whose yoke casings are welded to a large
rectangular steel main frame (1).   The rotors act in the manner of
keepers to these large electromagnets.   In this type of assembly,

the grooved plate matrix spends 50% of its lifetime within the field
of either C-magnet. Vertical feed pipes (8) are placed immediately
above the matrix and just within the point of entry to either field
system. Near the point of exit are wash pipes (13) to remove (12)
entrained non-magnetics middlings. In the field-free zones are a
number of scouring devices 14 to remove (10) the magnetic product.
The maximum field attainable in the gap between magnet and rotar is
approximately 2.0T. At the surface of the finest grooved plates
($\sim$2 mm groove pitch) the force density is of order 2×10^9 N m^{-3}.
This type of design is, therefore, well-suited to the concentration
of a fine-grained magnetic ore which constitutes a sizeable fraction
($\sim$50%) of the feed. It may, however, also be used for pre-concentration applications in low-grade tailings or for the upgrading of non-
magnetic minerals. Some typical mineral applications are summarized
in Table 1.

TABLE 1 (ref [9])

ACTIVITY	ORE APPLICATION
magnetic concentrates	haematite, pyrrhotite, siderite, ilmenite, ores of chromium, manga nese, tungsten, zinc, nickel, tantalum/niobium, molybdenum.
upgrading non-magnetic minerals	glass sand, apatite clay, talc, kaolin, feldspar, coal, fluorspar, nephelin, baryte, graphite, bauxite, cassiterite
preconcentration	tailings of uranium, gold, platinum, chromium and manganese.

The Jones separator is of crucial technological importance in the
beneficiation of both coarse (<5 mm) and fine-grained (<1 mm) feebly
magnetic iron ores.

Extensive experience with the DP317 (twin rotors of 3.17m
diameter) has indicated (averaged on a world basis) that an expected
throughput of around 120th^{-1} of crude ore (<1 mm particle size)
and with a grooved plate clearance gap of 2.5 mm [8]. With a 7600 h
annual running time this amounts to around 9.0×10^5 t y^{-1}. Process
costs include estimated capital costs of around $0.23 t^{-1}*based on
 years amortization at 10% (compound) interest, inclusive of
buildings, pumps, piping and installation. Manpower requirements
vary but may be reasonable estimated at 0.2 man-shifts per device |8|.

* 1984 estimates by the author ($\equiv$2.7 DM)

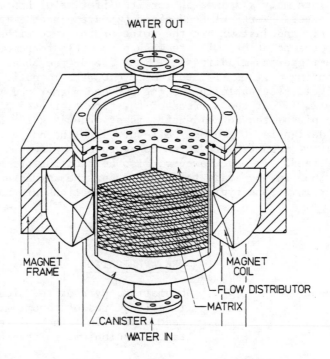

Fig. 4. Cutaway view of Kolm-Marston type of HGMS system
(Courtesy of Sala Magnetics Inc.)

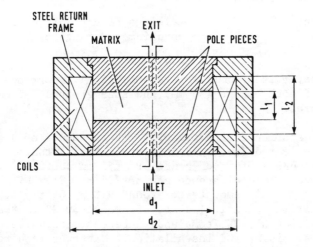

Fig. 5. Schematic of magnetic circuit of Marston-type iron-
bound solenoid (after Gerber [12]).

Power consumption is estimated to be between 0.5 and $0.7\,kWht^{-1}$ of feed (dry basis) [8]. Process water consumption for washing and scouring amounts to about 1.5 to $2.0\,m^3\,t^{-1}$ of feed (dry basis), about 80% of which can be recirculated.

§3. The Kolm-Marston 'High Gradient' Magnetic Separator

Around 1971 Kolm [10] formally introduced a modern magnetic separation device based upon the insertion of a filamentary, ferritic (corrosion-free) matrix into the working volume of a high-field solenoid. The novel aspect of this particular design is that, if sufficiently strong magnetic fields are generated by the solenoid, then strands of this filamentary matrix, whose axes are orthogonal to the field axis of the solenoid, are magnetised radially to saturation. Thus magnetic field gradients of order $2M_s/d$ are generated at the surfaces of the entire lengths of these fibres, where M_s is their saturation magnetisation and where d is the effective filament diameter. It follows that, under these circumstances, the field gradient at the surface of (an infinitely long) fibre of diameter 40 µm for which $M_s = 2.0$ T is of order $10^5\,Tm^{-1}$ and if the solenoid axial induction field is (say) 2.0 T then the corresponding magnetic force density at the fibre surface is greater than $10^{11}\,Nm^{-3}$. This is a full two orders of magnitude increase on the corresponding value for a Jones separator. It is clear, therefore that this type of separator is designed specially for problems involving the entrapment of small particles (<< 100 µm) which, in addition, are only paramagnetic. In fact, the impetus leading to the introduction of this device was such a problem, namely the removal of iron-stained microscopic titaniferous and micaceous particles from kaolin clays in Georgia, U.S.A.

The Kolm design concept was assisted greatly by the introduction by Marston [11] in 1971 of an elegant 'window-frame' design of the iron-bound solenoid (Fig. 4) which is now described in some detail.

In this system, the filamentary matrix is housed in a non-magnetic cylindrical canister which itself is housed in a pancake-shaped separation zone of approximate volume $\pi d_1^2 \ell_1/4$ (Fig. 5). Above and below this canister are cylindrical pole pieces of low-carbon steel which are magnetised by a close-packed water-cooled stack of copper coils (the coils are of square cross-section) wound around the perimeter of the canister, as shown. The cross-sectional area of the coils stack is $(d_2-d_1)\ell_2$. These coils are buried within a substantial (generally rectangular) steel flux return frame. In smaller systems, the feed and product streams respectively enter and exit through axial pipes as shown (Fig. 4) but in large-scale systems ($d_1 > 2.0$ m) there may be as many as 16 points of entry (or exit) distributed uniformly over the surfaces of the two pole pieces. One such large-scale separator, manufactured by Eriez Magnetics Inc., is illustrated photographically in Fig. 6. The device of Fig. 4 is cyclic which means that, after a number of canister volumes of feed stream have been processed, the

Fig. 6. Large-scale industrial HGMS system with 2.13 m bore (Courtesy of Eriez Magnetics).

performance (as measured in terms of recovery of magnetic material)
falls to an unacceptable level. The feed stream is then
interrupted (by valve closure), the magnetic field switched off,
the matrix washed and the field switched on again before resumpt-
ion of the process. The magnetic circuit has been described
in some considerable detail elsewhere [12] and only a brief
summary is given here.

A starting point in the design of a large-scale system of
the type shown in Fig. 5 is a target production rate, P, (kgs^{-1})
(of the feed material) and, from preliminary (laboratory) tests,
a suitable slurry density, ρ, (kg m^{-3}), stream velocity
V_o(ms^{-1}), and residence time τ(s) in the matrix, may be chosen.
The required matrix diameter, d_1, is given by $(4P/\pi\rho V_o)^{\frac{1}{2}}$ and the
matrix length ℓ_1 from $V_o\tau$. The required number of ampere-turns
(NI) for the copper coils may be estimated from Ampere's
circuital law. In other words, ignoring flux leakage, [12]
$(B_o/\mu_o)\ell_1 \simeq NI$ where I is the current in each hollow water-
filled copper tube, and B_o is the magnetic induction field
(in Tesla) parallel to the axis of the pancake-shaped separation
zone. The ampere-turns are generally supplied from a high
current – low voltage supply, the advantage of this being that
the coil system can be wound efficiently as an annular stack of
rectangular cross-section (of area $(d_2-d_1)\ell_2/2$) comprising square
cross-section hollow copper tubes containing water-carrying
channels. The copper filling-factor of these coils is around
0.75 and the current density is rated at about 3.0 x 10^6 Am^{-2}.
In a large-scale ($d_1 \simeq 2.1m$) device with a maximum operating
field of about 2.0 T a typical coil stack would be (say) an
11 x 28 (= 308 coils) rectangular array, each tube carrying a
current of around 3000 A and making a total of NI $\simeq 10^6$ ampere-
turns. From the resistivity of copper (with the coils running
slightly hot) of around 1.8 x 10^{-8} Ωm the total resistance (Z)
of the coils is around 0.05 Ω. The power supply to the magnet
(I^2Z), at full field, is therefore of order 0.42 MW [12]. These
figures are closely in line with those described in the
commercial literature [13] of Eriez Magnetics.

The major portion of the field in the canister is derived
from the magnetised pole pieces. A rough indication of the
contribution of each pole piece can be obtained from the end
field of a (semi-infinite) magnetised cylinder

$$B_{op} \simeq \frac{M_s}{2} [1 - \cos \{\tan^{-1} (\frac{d_1}{x})\}]$$

where x is the perpendicular distance from the centre of the
inner surface of the pole piece. The minor portion of the field
comes directly from the coils themselves. This latter
contribution may be expressed (at the geometric centre) in the

form [12]

$$B_{oc} = \frac{\mu_o NI}{(d_2 - d_1)} \ [\sinh^{-1} \ (\frac{d_2}{\ell_2}) - \sinh^{-1} \ (\frac{d_1}{\ell_2}) \ ]$$

In a 2.1 m diameter 2.0 T magnet it has been estimated [12] that of the total field, about 25% is derived from the coils at the geometric centre of the system. In the device outlined above $\ell_2 \simeq 2\ell_1$ while in other commercial designs a shorter coil stack ($\ell_1 \simeq \ell_2$) is preferred. Clearly, the details of design involve economic considerations involving a balance between increased copper (and operating) costs against higher capital costs for increased tonnages of low-carbon steel.

Apart from its robustness and reliability, the major attraction of this device is the efficient manner in which it can be scaled up from a simple laboratory instrument to a large industrial machine. This is not without advantages. An examination, for example, of published specifications of the SALA - HGMF [R] Low-Pressure Series of Magnetic Filters [14] shows (Fig. 7) that, over a wide range of HGMF filters the electrical power requirements (I^2Z) scale linearly with matrix diameter (d_1). Since P increases in proportion to $d_1{}^2$, there are clear advantages for large-scale designs of this type of system [15]. Against this, such systems have correspondingly large inductances with a calculated total value of L for the above - mentioned 2.1 m diameter system of around 1 H. With $Z \simeq 0.05 \ \Omega$ the relaxation time for switch-off of the magnetic field is of order 200 s. This can be reduced to around 60 s by dumping the current into secondary coils wound in opposition to and in a low inductance coupling with the main coils.

The production rate, P, of a cyclic HGMS system of the type shown in Fig. 6 can be expressed as a product of the mass flux $\rho \ V_o (\pi d_1{}^2)$ and of the duty cycle $n_o\tau/n_o\tau + n_r\tau + D$) where n_o is the number of equivalent canisters of the feed stream processed before the recovery, R, of the magnetic component falls to an unsatisfactory level. When this happens, the flow is halted by valve operation, and the matrix rinsed with $n_r(\simeq 1)$ canisters of wash water (at speed V_o and at full field) to remove all but magnetic material from the matrix. Then the field is removed, the matrix backwashed and the field restored. The time required for the entire sequence is known as the 'dead time' (D). Thus P can be expressed in terms of the separation zone geometry, as

$$P = \pi\rho \ d_1{}^2\ell_1 n_o \ / \ [n_o + n_r)\tau + D] \qquad (2)$$

The production rate for the product stream, if this is the non-magnetic component, is obtained simply by multiplying the RHS of equation (2) by (1-R).

It is evident that P depends crucially on the value of D especially since, in a well-designed system, ρ is increased to its maximum practical value which, in turn, reduces n_o to a value (say) of order 5. For large-scale conventional HGMS systems a value of D in the range 200 to 300 s is common.

A certain amount of criticism may therefore be levelled against the conventional large-scale cyclic 2.0 T system which operates in the fashion summarized above. In the processing of kaolin, for example, duty cycles in the range 40% to 50% can be encountered, with adverse consequences for the value of P, and thereby, for unit processing costs. A 2.1 m bore system, of the type shown in Fig. 6, costs ∿ $2.0 x 10^6 and manufacturers claim a processing rate of up to 31.5 ℓs^{-1} wet kaolin slurry. If the slurry contains 0.2 kg ℓ^{-1} of kaolin (dry basis) then, assuming amortization at 10% per annum and interest charges at 10% (compound) per annum the capital and interest charges process costs for this device are in the region of $1.4 t^{-1}. Operating power costs here are very high when the device operates at full field. From the above it can be seen that when kaolin is processed at full field the power

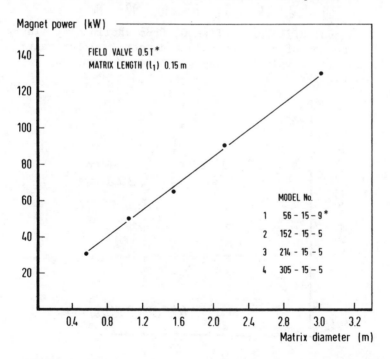

Fig. 7 linear dependence of magnet power (kW) on matrix diameter (m) in HGMS (ref. 1141).

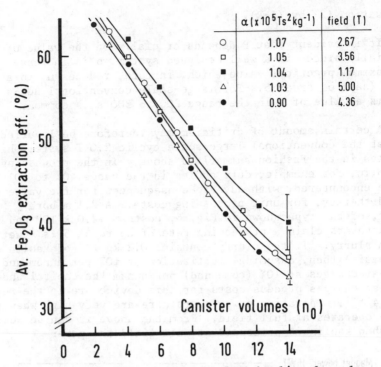

Fig. 8 Removal efficiency of Fe_2O_3 from kaolin clay slurry as a
function of canister volume number (n_o) for near-
identical values of the parameter α (courtesy of
J.H.P. Watson)

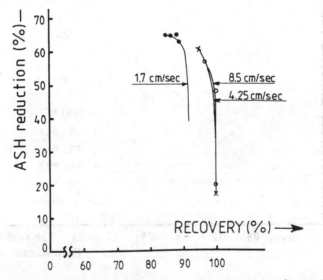

Fig. 9 ash reduction in coal slurries as a function of coal
recovery (courtesy of C.B.W. Kerkdijk)

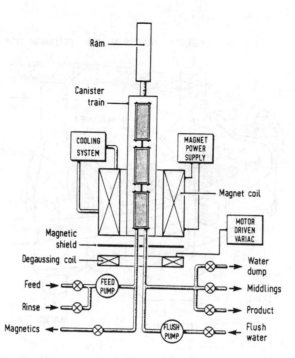

Fig. 10   reciprocating canister S/C HGMS system (courtesy of
P.W. Riley).

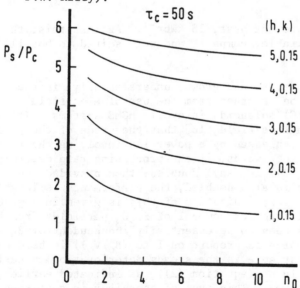

Fig. 11   production rate (P) comparison between superconducting
(S) and conventional (C) HGMS systems for various h
and k values (see text).

152

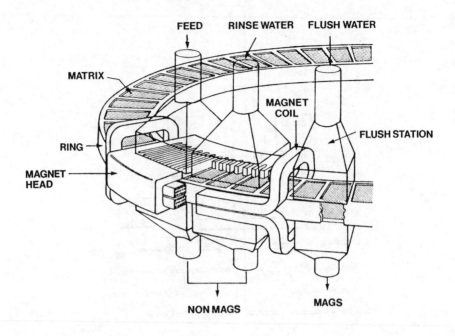

Fig. 12 schematic of SALA-HGMS TM CAROUSEL (courtesy of Sala
         Magnetic Inc)

requirements are, at best, 18 kWht^{-1}. Against this, the device
is robust, reliable, compact, and well-suited to hostile
environments.

    Recent interest has grown, understandably, in the possible
advantages to be accrued from the use of high-field superconducting (S/C) solenoids in cyclic HGMS systems. It is well
established in this field [16] that the grade of the product
stream can be expressed by a power law involving the ratio
(B_o/V_o). In other words, if the processing rate determined by
V_o, (equation (2)) is (say) doubled, then provided the
magnetic field is also doubled, the performance is largely
unchanged. A clear indication of this is given in Fig. 8 in
which the efficiency of removal of Fe_2O_3 particles from kaolin
clay slurry is seen to be essentially independent of B_o when the
parameter α (which is proportional to (B_o/V_o)) is held constant.
Also of interest here is the steady deterioration in performance
with increase in n_o (equation (2)), as indicated earlier, in
more general terms. This type of behaviour is confirmed by
some recent work by Kerkdijk et. al. [17] on coal slurries. In
Fig. 9 the recovery of coal is seen to be near-independent of a
simultaneous doubling of field and a flow rate.

Fig. 13 SALA-HGMS^{TM} CAROUSEL (courtesy of Sala Magnetic Inc.)

Having said all this, it must be pointed out that the
direct replacement of a conventional solenoid by a superconducting
one is unlikely to produce any real benefits either in performance
or in cost. A large-bore (say 1.0 m) S/C solenoid would have,
normally, a dead-time in the region of 500 to 600 s. (A
laboratory-scale S/C high-field solenoid with a fast ($\sim$ 34 s)
rise and fall ($\sim$ 30 s) time is currently being marketed by Eriez
Magnetic but this is, in comparative terms, expensive, costing
approximately \$300,000 for a separation zone of volume of
approximately 4.0 ℓ). Therefore, all of the benefits to P in
equation (2) from an increase in V_o (i.e. a reduction in τ)
allowed by the high S/C field are offset by the very large value
of dead-time. The only effective way in which S/C magnets can
be employed is in some truly continuous system (such as the
cryogenic drum separator [18] or in a cyclic system in which D
is, somehow, reduced (preferably to values much lower than its
conventional counterpart). A scheme, first proposed by Riley
and Hocking [19], from original designs and patents of

Z.J.J. Stekley, shows some considerable promise in this area. The scheme referred to (Fig. 10) is the reciprocating canister (RC) system. Here, as the diagram indicates, the S/C field remains permanently switched on (for example, locked in the 'persistent' mode). When n_o reaches a valve (Fig. 8) at which the performance is unacceptable, a value is closed, the canister rinsed and then removed by a hydraulic ram from the solenoid. As it leaves the solenoid it is replaced by canister with a fresh matrix. These canisters are, in fact, part of a linear canister train designed to minimise the hydraulic force necessary for canister removal. A more detailed analysis is given by this author elsewhere [16].

There are a number of advantages to use of the RC system. As Fig. 10 indicates, the withdrawn canister can be fully shielded magnetically, and de-gaussed during washing. The wash period can be extended to a time approaching the residence time $(n_o \tau_{S/C})$ of the live canister in the field (without prejudice to the duty cycle). Most important, however, is the fact that D can be reduced to a value simply equal to the withdrawal time of the spent canister from its position in centre field. Watson [20] has estimated that a value of $D \simeq 10$ s is achievable in large-bore high field S/C systems.

Fig. 11 indicates a production rate (P) comparison between a conventional (c) and a S/C system (s) in which $D_c = 200$ s and $D_s = 30$ s. If τ_c is fixed at (say) 50 s then $\tau_s = \tau_s/h$ where $h(= B_s/B_c)$ is the field advantage of the S/C solenoid. Here, $k = D_s/D_c$. Clearly, when the matrix is washed out after only a few equivalent canisters have been produced (i.e. $n_o \ll 10$) the production rate advantage P_s/P_c is better even than the field advantage, h, on account of the engineering advantage (i.e. low D - value) of the RC system.

RC systems, do not, as yet, offer clear <u>economic</u> advantages over the conventional system of Fig. 6. Watson [20] has recently described a 0.30 m diameter, 5.0 T, RC system currently under construction in the U.K. Here, the total capital cost is approximately $400,000, inclusive of the (20kW) refrigerator and of the canister train and hydraulic ram. Under the best of circumstances (with $V_\rho = 50$ mms^{-1} and with a duty cycle of about 80%) this unit could be expected to process about 25,000 t of high-grade kaolin clay per annum. At 10% amortisation and 10% (compound) rate of interest on the capital investment this produces capital depreciation and interest costs as high as $4 t^{-1}. However, it must be pointed out that the operating power costs here are almost entirely, those of the refrigerator (20 kW). Moreover, such a system can be automated fully. The commercial prospects for RC systems will only become clear if the

reciprocation principle can be demonstrated successfully on
systems of diameters approaching 1.0 m and if the capital cost of
such a system compares reasonably well with the conventional HGMS
of Fig. 6.

Current technological and economic considerations set an
absolute upper limit on S/C large-bore solenoids of around 8.0 T.
If this is reduced to (say) 5.0 T the economics have probably,
their optimum field value. This is also a reasonable upper
field limit for removal of the matrix in a distortion-free manner.

The RC system has only been tested in principle (satisfactorily) in a small-bore system [19]. However, numerous recent
data are available in the literature on high-field HGMS
experiments in mineral processing. In addition to showing
excellent brightness gains in fine kaolin clay, Gillet and
Houot [21] have recently shown favourable mineral processing
performance figures for a S/C HGMS system compared with a
conventional wet high intensity magnetic separation (WHIMS) (of
unspecified manufacture). In the upgrading of nephelin syenite
raw feed (a) ground to sizes between 100 μm and 50 μm and (b)
comprising the −40 μm fraction were subjected to separation tests.
Results confirm that about 98% of the iron and titanium impurities
are removed in a single pass at 9 mm s^{-1} in a 5.0 T S/C solenoid.
The concentration of itabirite ore fines is also shown (Table 2)
to compare favourably with a Jones separator and with indirect
flotation.

Watson and co-workers [22] have used similar techniques in
the preconcentration of leached South African mine residues of
gold and uranium oxide. Here, they have found that 54.4% of the
total mass of uranium occurs in the −25 μm slimes. These they
have subjected to wet HGMS treatment at magnetic fields of up to
7.0 T. They have found that, if the small percentage of
ferromagnetic contamination in the ore is scalped off, (using a
paramagnetic matrix), the uranium ore fines can be upgraded by
factors of about 3 in a single pass (Table 3).

HGMS systems can hold, usually, a volume of paramagnetic
material roughly equal to the matrix volume before the
performance becomes unsatisfactory. As such the cyclic systems
outlined above are unsuitable for the processing of wet mineral
slurries in which the magnetic component is the valuable
product. Sala Magnetics Inc. have produced a range (two of which
are summarized in Table 4) of continuous HGMS devices based on a
rotary principle like the Jones separator of Fig. 3. This is the
SALA-HGMS^{TM} CAROUSEL which is illustrated schematically in
Fig. 12 and photographically in Fig. 13. Unlike the Jones
separator, the Carousel has no central rotary shaft and is,

TABLE 2   (After Gillet et.al. [21]

PRODUCTS	WEIGHTS (%)	Fe(%)	RECOVERY Fe(%)	TEST CONDITIONS
Fe concentrate	68.0	63.44	99.3	magnetic field:2T
				feed:20% solids
tails	32.0	1.03	0.7	V_0 : 1.5 cms^{-1}
total field	100.0	46.43	100	matrix:40 cm length
				(steel wool)
				particle:< 37 μm
				size
Fe concentrate	49.65	67.32	72.96	3 passes on
middlings	10.63	23.12	5.36	Matrix:grooved plates
tails	39.72	24.99	21.68	of gap 0.8 mm
feed	100.0	45.81	100.0	V_0: 1.5 cm s^{-1}
				Magnetic; 1.5 → 1.9 T
				field
				particle: <37 μm
				size
Fe concentrate	61.39	67.26	92.82	indirect flotation
tails	38.61	8.27	7.18	(1 roughing flotation,
feed	100.0	44.48	100.0	2 froth cleaning)
				particle size:<44 μm

TABLE 3   (after Watson et.al. [22]

TEST NO.	B (T)	B/V$_0$ (arb. Units)	Scalped feed	Uranium upgrading	Uranium recovery (%)
1	2	4.7	no	2.0	43
2	4	9.7	no	2.4	36
3	5	7.7	no	1.3	31
4	5	4.1	no	1.7	28
5	4	10.7	yes	3.3	51
6	6	8.0	yes	3.3	49
7	7	6.0	yes	3.8	61
8	6.2	27.1	yes	3.3	49

instead, chain driven. As Fig. 11 indicates the magnet here is a
localized 'saddle' magnet whose field axis, unlike the Jones
devices, is vertical. The matrix, (like its cyclic Kolm-Marston
counterpart in Fig. 4) is, typically, a compartmented stack of
expanded metal screens. On the largest model – the – 480 –,
the manufacturers claim processing rates as high as 600 th^{-1}. The
range of applications of these devices is essentially the same
as that described for the Jones separator in Table 1. This
device, however, is designed to operate effectively on much
smaller particle sizes. The price paid is that of considerably
higher power consumption per tonne of mineral processed. For
example the DP 317 processes 120 th^{-1} at about 0.5 - 0.7 kWht^{-1}.
The Carousel 350 processes 100 th^{-1} per head at a power consumption
of 250 kW – that is, at 2.5 kWht^{-1}.

In summary, it can be said that Jones and Kolm-Marston
separators have both been applied successfully to wet mineral
processing in the same three areas of activity described in
Table 1. In general terms the latter are more efficient in areas
in which the particle size is very small ($\ll$ 100 μm) and where
the magnetic volume susceptibility is low. Recently, the Kolm –
Marston continuous (Carousel) separators have been applied with
moderate success to problems of dry magnetic separation [24], but
it will be some time before an informed appraisal can be made of
the technological significance of this.

It should be noted, finally, that as ∇H increases to higher
and higher values a corresponding reduction is to be expected
in the effective range of the magnetic traction force. In the
Jones and Kolm-Marston separators these ranges are of order
1.0 mm and 0.1 mm, respectively. Open gradient separation
devices, however, are characterized by much smaller values of
field gradient and, consequently, of force density. This is
partially compensated, however, by much larger values of the
force range, as will be seen in the following.

§3 Particle Deflection Devices

In the particle deflection mode of separation – often
referred to as open gradient magnetic separation (OGMS) – the
separation of components is obtained by subjecting mixed (wet or
dry) particle streams to appropriately designed non-uniform
macroscopic magnetic fields. Particles in the feed stream are
deflected selectively in these fields to distinct and separate
geometrical points of collection according to size, density
(ρ) and magnetic volume susceptibility (χ).

In the laboratory, this concept is not new to mineral
processing engineers. The Frantz Isodynamic Separator ® works
precisely on this principle. Here, mixed powders are compelled by

TABLE 4   (ref [23])

MODEL NUMBER	120 A-B	480-A.B-C.D
width, breadth, height (m)	4, 2, 2	8, 8, 4.5
potential number of magnets on ring	2	4
max. capacity/head (t)	3	200
field rating (T)	2.0	2.0
electrical power/magnet (KW)	250	300
basic weight (t)	4.5	18

TABLE 5   (after Kopp [31]).

mean particle size (μm)	V_o (mms^{-1})	λ (mm)	number of collisions	lateral displacement (mm)
50	32	0.5	50	2.5
100	45	1.5	17	2.0
200	60	2.8	9	2.0
500	100	5	5	1.9
1000	140	7	4	2.1

electro-mechanical agitation to proceed down an inclined chute whose path is along the axis of an isodynamic magnet. The chute is tilted sideways so that the magnetic traction force (expressed by equation (1) is opposed by a component of the gravitational pull. Both forces are proportional to particle volume so that particles are deflected laterally (to either side of a splitter) as they progress down the chute in a fashion which is independent of particle size and which depends only on mass susceptibility (χ/ρ).

A variety of non-uniform magnetic fields systems have been used to promote this type of separation on a large scale, [25]. Of particular note here is the annular quadrupole magnet system used first by Kolm [26] and, later, by Cohen and Good [27].

Most attention has focussed, in recent years, on the use of S/C linear multipole systems of the sort first suggested by Shoenert et. al. [28]. In the last four years a simplified version of this design, namely a S/C 'split-pair' linear quadrupole has received the greatest scientific attention in pilot studies in the field of dry mineral processing [29,30]. Here, (Fig. 14) powdered minerals fall freely under gravity in a thin annular curtain in close proximity to the outer wall of a cylindrical cryostat. In the vicinity of the coils the particles enter a (predominantly radial) field gradient with a force density pattern of the type shown in Fig. 15. In dry separation the particles are subjected to lateral deflection proportional to $(\chi/\rho)^2$ [25]. For particles of sizes > 0.5 mm these deflections and the consequent separation process is approximately isodynamic [25]. As Fig. 14 illustrates, an annular splitter (or series of splitters) just below the separation zone allows continuous collection of dry mineral particles in two (or more) fractions.

Although simple in its design concept, this mode of separation has, associated with it, some fundamental problems relating mainly to inter-particle collisions. Kopp [31], in a detailed analysis of this problem, has shown that, in a dense falling stream of particles, of cross sectional area A, containing n particles/unit volume, (each of volume v_p), then at some vertical position at which the stream velocity is v, the equilibrium mass flux, P, will be

$$P = An\, v_p \rho v \qquad\qquad (3)$$

Since P, A and ρ are constants, then

$nv_p v = V_o$ is a constant, known as the 'characteristic feed velocity'. If this stream falls from a hopper whose aperture

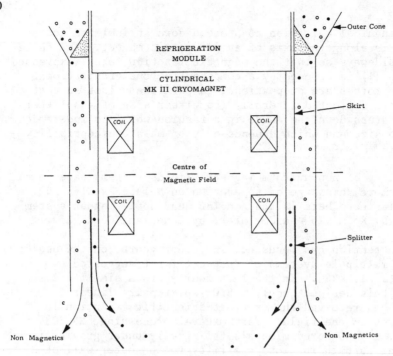

Fig. 14   split-pair OGMS system (courtesy of E.H. Roux [24])

Fig. 15   radial force density profile of S/C split pair (in
T^2 cm^{-1})(courtesy of E.H. Roux [24])

is very much larger than the particle radius, b, then

$$V_o \div (2gb)^{\frac{1}{2}}$$

Kopp argues that, as particles are pulled radially inwards by the magnetic traction forces of Fig. 1.5, their 'mean free path', λ, between collision will be

$$\lambda = (n\sigma)^{-1}$$

where $\sigma \div 4\pi b^2$ is the 'scattering cross-section'. He therefore concludes that

$$\lambda \sim bv/3V_o$$

which means that particles pulled inwards through a curtain of thickness t experience n_c collisions, where

$$n_c \sim t/\lambda .$$

Table 5 shows tabulated values of n_c for various values of b for a certain of thickness 25 mm. These calculations seem to confirm recent dry OGMS measurements of Roux et. al. [24] on powdered apatite in which the performance is found to decrease rapidly as particle diameter [26] falls below 100 μm.

Equation (3) is also useful for an estimate of processing rate P (per metre length of curtain) of OGMS systems. Male [32] has quoted experimental large-scale values of (nv_p) of ~ 0.04. If v is taken as 1 ms^{-1} then for a mineral of density 3 x 10^3 kg m^{-3} and for t = 25 mm,

$$F \sim 10 \text{ th}^{-1}\text{m}^{-1} .$$

This is consistent with projected estimates of performance [24] of OGMS systems for the processing of dry pyroxenite ores.

Prototype OGMS Systems

To date, only published data on dry OGMS is available for small cylindrical cryostat systems (of the type shown in Fig. 14) with outer cryostat wall diameters of the order of 0.4 m or less [29,30]. Two of the leading producers in this area are Cryogenic Consultants Ltd. (CCL) and Oxford Instruments Ltd. Both offer refrigerated systems but some variance exists in their design philosphies. The (Mark 2 & 3) S/C split pair magnets of CCL use only small quantities of liquid helium to maintain the coils below their critical temperatures. The cooling technique employed here is a 2-stage continuous recompression of helium gas in addition to refrigeration by expansion and heat exchange. This type of system therefore requires no external liquid helium supplies. However, in the event of power supply or refrigerator (compressor) failures magnet quenching is almost instantaneous. The delays involved in re-cooling a magnet ($\sim 1\frac{1}{2}$ days) after this type of failure may well probe unacceptable in an industrial environment.

Oxford Instruments have designed their cryostat on more conservative principles [33]. Here the magnet is maintained at 4.2 K by boil-off from a substantial liquid helium reservoir but a small closed-cycle refrigerator is incorporated to minimise this process as well as that of liquid nitrogen losses. The refrigerator comprises a 2-stage expansion engine providing two stations external to the cryostat tail with temperatures of << 77 K and < 15 K. The former is coupled to the liquid nitrogen reservoir in such a manner as to prevent boiling. The 15 K station reduces radiation heating of the helium reservoir to a point at which liquid helium boil-off is less than 0.01 ℓh^{-1} in the persistent mode. This magnet system has operated continuously at full field in the persistent mode for about 6 months at the Royal School of Mines [33] at the time of writing and has been shown to maintain full field for a period of at least 12 hours in the event of power or compresser failure.

Now, as Fig. 15 has already indicated, the magnetic force field of the cylindrical split-pair magnet diminishes rapidly with radial distance from the cryostat wall. This force field distribution increases only very slowly with increasing cryostat diameter (and eventually saturating) so that, unlike for example HGMS there is no economic benefit in scaling up this cylindrical geometry for industrial scale OGMS. Good [34] has circumvented this particular scaling problem by abandoning the cylindrical geometry completely in favour of a linear geometry where the split pair is replaced by a 'hairpin' coil stack. An added bonus in this linear format is the comparative ease with which a 'straight-edge' splitter may be positioned to optimize grade and recovery in the separation process compared with the extreme practical problems associated with annular splitters in the equivalent cylindrical geometry. A 3-metre long linear OGMS system of this sort has recently been developed by CCL for dry OGMS processing of phosphate ores in South Africa [34]. The separator can be utilized along almost its entire length on both sides of its axis which means that it can dry process a mineral curtain of approximately 6 m length. This means, at a capital cost of around $670,000, a total processing rate (P) of around 60 th^{-1}. On the same basis as for other separation systems described earlier, this means at 1984 prices, capital interest and depreciation unit costs of only $0.39 t^{-1}. Moreover, as Roux et. al. [24] have recently discussed in detail, dry

separation offers significant cost advantages over wet ore
processing in terms of the elimination of the use of water,
pumps, pipelines, slurries dams, thickness, filters, driers,
flotation reagents, flocculants and so on, leading to projected
capital advantages over conventional flotation of 20%.

§4   Projected Future Developments

The highly original separation systems shown, for example,in
Figs.10 and 14 may have a profound influence on the development of
magnetic separation technology in the coming decade. By no means,
however,do these and the other systems described here give a
comprehensive picture of future developments. For example, since
the first patent was launched by Ries et.al. [35]in 1978 for a S/C
magnetic drum separator, considerable successful industrial
development has been undertaken in this area by KHD Humboldt
Wedag AG but propriety dictates that no details can be released
at this stage [36].

Acknowledgements

The author is grateful to the various authors and commercial
organizations for permission to reproduce photographs, line
diagrams and results.

References

1. Dean, R.S. and C.W. Davis. Magnetic Separation of Ores.
    (Bureau of Mines Bulletin 425, U.S. Dept. of the Interion,
    1941).
2. Taggart, A.F. Elements of Ore Dressing. (New York, John
    Wiley, 1951).
3. Pryor, E.J. Mineral Processing. (New York, Elsevier, 1965).
4. Parker, M.R. The Physics of Magnetic Separation. Contemp.
    Phys. (1977) 279-306.
5. De Vaney, F.D. New Developments in the Magnetic Concentration
    of Iron Ores. 5th Int. Minerals Processing Cong., London
    (1960) 745-749.
6. Jones, G.H. British Patent No. 768, 451 (1955).
7. Jones, G.H. Wet Magnetic Separator for Feebly Magnetic
    Minerals. Proc. 7th Int. Minerals. Processing (org. (1964)
    717-732.
8. Bartnik, J.A., H.-D. Wasmuth and W.H. Zabel. Production of
    Coarse Grained Iron Ore Sinter Feed Using The Jones WHIMS.
    Sonderdruck aus Zeitschrift. Auf. Technik. Sept. (1982)
    S490-497.
9. K.H.D. Industrieanlagen A.G. Publication No. 4-720e.

164

10. Kolm, H.H. U.S. Patent No. 3567026 (1971).
11. Marston, P.G. U.S. Patent No. 3627678 (1971).
12. Gerber, R. Some Aspects of the Present Status of HGMS.
     I.E.E.E. Trans.on Magn. vol MAG-18 (1982) 812-816.
13. Eriez Magnetics Publication No. OTB-510A (1981)
14. Sala Magnetics Inc. Bulletin No. 24200107-7715GB (1977).
15. Robinson, G.Y. and P.G. Marston. Application of High
     Gradient Magnetic Separation to Clay Beneficiation. Proc.
     of Int. Clay Conference, Mexico City, July 1985.
16. Parker, M.R. Use of Superconducting Magnets in Magnetic
     Separation. J. de Physique. Vol. Colloque C1, supplement
     au No. 1 (1984) C1-753-C1-758.
17. Van Driel, C.P., C.B.W. Kerkdijk, H.R. Segal and J.
     Sikkenga. Coal Cleaning by HGMS. VMF-Stork, FDO and Holec
     Netherland Publication. August 1983.
18. See concluding section.
19. Riley, P.W. and D. Hocking. A Reciprocating Canister
     Superconducting Magnetic Separator. I.E.E.E. Trans. on
     Magn. Vol. MAG-17 (1981) 3299-3301.
20. Watson, J.H.P. Seminar, Univ. of Nijmegen, July 11 (1983).
21. Gillet, G., R. Houot and G. Leschevin. Magnetic Separation
     with a Superconducting Solenoid Type Magnetic Separator
     Applied to Mineral Processing. Proc. Int. Symp. on
     Electrical and Magnetic Separation and Filtration Technology,
     Antwerp (1984) 97-108.
22. Watson, J.H.P., D. Rassi, D. Prothero and R. Potts. High
     Gradient Magnetic Separation of Uranium Oxide and Gold from
     Mine Residues. Proc. Int. Symp. on Electrical and Magnetic
     Separation and Filtration Technology, Antwerp (1984) 93-96.
23. Sala Magnetics Inc. Bulletin No. S79-0001 (1979).
24. Roux, E.H., J.G. Goodey, E.F. Wepener and K.R. Hodierne.
     Industrial Scale Dry Beneficiation of Phosphate-Bearing
     Pyroxenite Ore. Proc. of Mintek 50, Randburg, S. Africa
     (1984).
25. Parker, M.R. Recent Developments in High Field Magnetic
     Separation. Proc. Int. Symp. on Electrical and Magnetic
     Separation and Filtration Technology, Antwerp (1984)1-13.
26. Maxwell, E. Magnetic Separation. The Prospects for
     Superconductivity. Cryogenics (1975) 179-184.
27. Cohen, H.E. and J.A. Good. Principles, Design and
     Performance of a Superconducting Magnet System for Mineral
     Separation in Magnetic Fields of High Intensity. 11th Int.
     Mineral. Proc. Conf. Cagliori (1975) 773-793.
28. Schonert, K., A. Supp and H. Dorr, 12th Int. Minerals
     Processing Cong. Sao Paulo, Brazil (1977) paper No. 1,
     Meeting No. 4.
29. Collan, H.K., M.A. Kokkala, T. Meinander and D.E. Toikka,
     Superconducting Open-Gradient Magnetic Separator. Trans.
     Inst. Min. Metall. Vol. 91 (1982) C5-C8.

30. Kopp, J. and J.A. Good. The Physics of High Intensity Dry Magnetic Separation. Vol. MAG-18 (1982) 833-835.
31. Kopp, J. The Physics of "Falling Curtain" Dry Magnetic Separation. Int.J. of Min.Proc. Vol. 10 (1983) 297-308.
32. Male, S.F. (private comm.)
33. Cohen, H.E., S. Roberts and N. Kerley. A New Superconducting Magnet System for High Intensity Magnetic Separation. Proc. Int.Symp. on Electrical and Magnetic Separation and Filtration Technology. Antwerp (1984) 85-92.
34. Good, J.A. and K. White. The Design and Construction of a Superconducting Magnet for Production Scale Dry Separation of Minerals. Proc. Mintek 50, Randburg, S. Africa (1984).
35. Ries, G., K.-P. Jungst, S. Forster, W. Lehmann and K.-H. Unkelbach, West German Patent No. 2650540 (1978).
36. Full details will be presented by K.H.D. Humboldt Wedag A.G. at the Int. Mineral Proc. Congress, June 1985 (Cannes).

THE METHODOLOGY OF FROTH FLOTATION TESTWORK

R.J. Gochin, M.R. Smith

Mineral Technology Section, Imperial College, London.

ABSTRACT

The application, procedures and criticsms of all scales of froth flotation testwork are reviewed ranging from single mineral studies in the laboratory to pilot plant trials.

The importance of the initial stage of defining the objectives and evaluation criteria of testwork together with a considered experimental plan is emphasised. Techniques of experimental design and statistical evaluation of results are presented such as significance testing, ANOVA, factorial design and optimisation.

Practical aspects of the popular batch flotation tests are discussed in detail especially with respect to process flowsheet development for a particular ore. Correct sampling and a prior comprehensive mineralogical study of the ore are considered to be essential.

Finally, the mathematical modelling of the kinetics of the froth flotation process is described with particular relevance to "scale-up" and equipment selection.

1. INTRODUCTION

Investigations of froth flotation can vary from the testing of novel reagents to float specific minerals to the determination of the type, size and number of flotation cells required for a full scale plant. The common investigations are related to the development or application of a froth flotation process to a particular ore and evaluating the potential benefits of altered froth flotation conditions upon an existing plant (1).

This paper reviews the types and applications of the test procedures that are available. In addition, the importance and methods of designing and evaluating testwork are discussed in order that a great deal of effort does not culminate in inconclusive results or, worse, misleading interpretation of results (2).

2. OBJECTIVES AND EVALUATION CRITERIA

The first prerequisite for testwork is a clear definition of the objective and, thereby, the criterion by which the testwork is going to be judged. This criterion may not be a simple technical parameter such as grade or recovery but may be of an economic nature involving the calculation of the "value" or incremental increase in value of a concentrate and the "cost" or incremental cost of producing it. These considerations are important since they should permit a statement of the quantity, i.e. time and cost of testwork that can be justified plus a definition of the quantity and accuracy of testwork that is likely to be required in order to produce a conclusive result. One "hidden" cost of testwork can be that it protracts the development of a deposit or a change in practice. In some cases a particular type of testwork may offer great potential benefit; in others the cost may not be justified when compared to the potential benefit.

3. MINERAL APPRAISAL

In the majority of cases testwork will employ samples of natural materials with the objective of separating one or more distinct phases. In this context a separate discussion and emphasis upon the importance of mineralogical studies is necessary when reviewing laboratory froth flotation testwork (3,4,5,6).

3.1. Sampling

It is essential that the sample used in the testwork is mineralogically representative of the orebody and chemical analysis gives only indirect information in this respect. A mineralogical description of the ore must be made to include the identity of minerals present, their relative abundance, `modal analysis´, and information concerning the grain size distribution of each mineral, their intergrowths and other textural relationships.

3.2. Chemical assay.

Fortunately, a number of excellent instrumental technqiues now exist for the assay of most elements, for example; atomic adsorption spectroscopy (A.A.) for most important base and precious metals e.g. Ag, Au, Cu, Pb, Zn, Fe; and plasma emission spectroscopy for the more refractory metals, alkali earths and non

metals e.g. Sn, W, P, S, Ni, Co, and very low concentrations of
precious metals e.g. Au. X-ray fluorescence analysis provides a
very rapid method of analysing dry powders for elements between Na
and U. Only carbon (carbonate) and fluorine continue to pose a
problem requiring traditional "wet assay" technqiues and electron
probe equipment respectively. A very useful technique in ore
evaluation is the semi-quantitative X-ray fluorescence scan which
yields assay values for all elements as percent weight in orders of
magnitude. This greatly assists the mineralogical study in that
"host minerals" for all elements of significance detected must be
found. In addition, it should be possible to fully reconcile the
chemical assay with the "modal analysis" determined
microscopically.

The chemical assay itself, however, does not give any direct
information concerning the distribution of elements between various
minerals. This can be extremely important in that it will
determine the maximum grade of concentrate and recovery of value
that can be achieved by any given froth flotation concentration
process.

3.3. Mineralogy

The most important tools are still the optical microscope
using both reflected and transmitted light and the binoccular or
stereo scopic microscope (7) . It is unlikely that a flotation
engineer will be sufficiently trained and experienced to identify a
wide range of minerals microscopically . However, it is possible
to learn to identify the relatively small number that will occur in
any one given ore sample. Examination of sections of the ore and
identification of minerals will clearly indicate the necessary
general flotation conditions, the degree of comminution necessary
to achieve liberation and the likely limits that will be imposed
upon concentrates assays and recoveries by mineral intergrowth.

In some cases the valuable or detrimental elements are present
as sub-microscopic inclusions or isomorphous substitutions. The
distribution of these elements can now be determined using the
quantitative scanning electron microscope (microanalyser) (8). The
same machine can also automatically produce a modal analysis and a
mineral grain size distribution.

Heavy liquids and laboratory concentration devices such as the
`Super-panner´ and Franz Isodynamic separator can be useful to
isolate or concentrate certain fractions of the ore and particles
may even be hand sorted under the stereo-scopic microscope. These
fractions can be submitted to chemical or X-ray diffraction
analysis for mineral identification.

In conclusion, it must be stressed that a thorough
mineralogical study is a necessary prerequiste to flotation

testwork rather than just a regressive stage of attempting to determine the cause of unsuccessful testwork. Neither should this form of study be limited to exploration ore samples because very useful information can be gained by examining the products of froth flotation separation at all scales, especially if the products are divided into closely sized fractions (9).

4. TYPES OF FLOTATION TESTWORK

The various forms of flotation testwork are summarised below;

1. Small scale techniques.

2. Batch flotation tests during which only one variable at a time is changed.

3. Batch flotation tests during which a number of timed samples of froth are collected.

4. Locked cycle tests during which a number of batch flotation cells are used but continuous operation is simulated by the recirculation of intermediate products and especially water.

5. Continuous flotation testwork, pilot plant.

4.1. Small scale tests.

These tests (10), such as, contact angle studies, bubble pick-up tests, Hallimond tube (11), surfactant adsorption studies,(12,13) two liquid flotation, electrophoresis and electrochemical techniques (14) employ very small samples (< 1 g) or single mineral particles and offer the advantages of low cost, convenient methods of determining the general conditions under which minerals may be rendered floatable. They are particularly useful for the study of newly developed reagents (16) and unusual minerals (15). However, the limitations of these tests should be appreciated. They can take little or no account of (i) the physical process of flotation, i.e. such effects as particle size and bubble size, agitation and froth structure or (ii) of other minerals that would be present or soluble ions derived from the ore or grinding process. Furthermore, "pure" mineral specimens are often used that may have been obtained from deposits other than those for which the process is being tested and which are often variable in their properties owing to ionic substitution. Therefore, it is very important to characterise those specimens that are used.

4.2. Batch Flotation Tests - "One-variable-at-a time"

These tests usually employ between 500 and 1000 g of sample of the ore under investigation and, therefore, overcome many of the

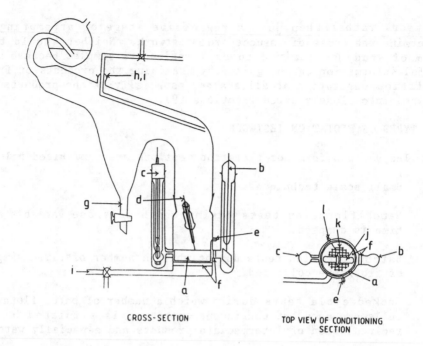

CROSS-SECTION

TOP VIEW OF CONDITIONING
SECTION

Figure 1 illustrates the equipment used for controlling the
electrical potential during conditioning as a means of obtaining
selective flotation of sulphides.

shortcomings of small scale tests. They are extremely useful for
the development of a general froth flotation process flowsheet and,
if correctly designed, for comparative tests to evaluate the
improvements that might be obtained by changing flotation practice.
The problems of directly relating the results obtained to plant
practice (scale-up) are dealt with separately.

      The following factors and controls should be considered before
embarking upon batch flotation testwork if meaningful results are
to be obtained (17,18).

        (a)   Sampling
        (b)   Age and alteration of sample
        (c)   Froth removal procedure
        (d)   Grinding of sample
        (e)   Water quality
        (f)   pH
        (g)   Pulp density
        (h)   Aeration rate and agitation
        (i)   Temperature

(a) <u>Sampling</u>. It is important that the sample be mineralogically, not just chemically, representative of the orebody. Since the material for this testwork is often drill core, it may be necessary to test various different recognised types of ore depending upon the results of the exploration programme. It is a useful practice to split drill core using part for testwork and retaining the other part for further mineralogical study if this is found to be necessary. The testing of composites of drill core is only useful if it is envisaged that the eventual mining schedule will be able to blend the ore in the same manner.

(b) <u>Age and alteration</u>. Many minerals degrade when exposed to the atmosphere, notably the sulphide minerals. It is thus often necessary to ship such samples dry, in sealed containers, at as large a lump size as practical and, maybe, in an inert atmosphere. Alteration of minerals and reagents may occur in the test pulp itself resulting in the release or generation of interferring species. This can be a particular problem of batch testing and especially of locked-cycle testing where the durations of grinding, flotation and solid-liquid separation steps are very much longer than those in a plant.

(c) <u>Comminution</u>. Grinding affects the froth flotation test in three ways;

      (i)   control of the degree of liberation

      (ii) control of the range of particle sizes present

      (iii) contamination of mineral surfaces (by oxidation or other chemical reaction and smeering of one mineral on another).

In the case where modifications to existing operations are being tested it is best to take the sample from the comminution circuit of the plant under investigation. However, if the excercise is one of process development there are two ways in which the problem of relating the laboratory grind to plant practice can be partly solved. The use of laboratory batch rod mills to prepare the sample for the flotation test is common as it has been shown that the size analysis of the product closely approaches that from an industrial ball mill operating in closed circuit (19). In addition, it is often very informative to examine microscopically closely sized fractions of the products of flotation testwork, i.e. the tailings and concentrate. This will indicate problems caused by inadequate liberation plus other particle size effects such as slime-gangue entrainment and particles both too coarse and too fine for efficient separation (20). In this way, the optimum particle size range can be determined. For comparative testwork it is only important that the samples be prepared by identical grinding technqiues.

A technique is available for establishing the perfect separation by froth flotation of an ore sample in a given state of liberation. The method is called "release analysis" and results in a "release curve" which is equivalent to a concentrate grade versus recovery of value curve for a perfect separation (21). In this respect, release analysis for flotation is analogous to heavy liquid separation for gravity concentration. The "release curve" has been shown to be very largely independent of operator, cell type and flotation conditions and thus represents a measure of the state of liberation (or release) of the ore. The great advantage of the technique when developed was that it overcame the problems of bias. The disadvantages of the technique are that the procedure involves many batch tests, although a simplified procedure has since been proposed (22).

(d) <u>Froth Removal</u>. Recent studies upon processes occuring within a froth have shown quite clearly that a concentration gradient of the floating mineral exists (23). On many plants, means of regulating pulp level, or more importantly, the froth depth have been installed in order to control concentrate grade and recovery (24). For the same reasons, the results of batch flotation testwork have been found to be sensitive to the method of froth removal and pulp level control in the cell. Since froth removal and pulp level control have been traditionally manual operations by the technician, these effects became the major sources of so-called "operator error". It was thus very important to employ the same technician for each one of a series of comparative batch flotation tests and to standarize the removal procedure.

This problem can now be largely overcome by the use of automated laboratory flotation machines such as the "Leeds cell", see Figure 2. With this cell, pulp level and pH are adjusted automatically and froth removal is independent of the operator. Aeration rate and impeller speed can also be independently measured and controlled (25,26). All these modifications result in excellent reproducibility of results which is the most important criterion for comparative testwork.

(e) <u>pH regulation</u>. Several pH regulators are available e.g. lime, caustic soda, sodium carbonate, sulphuric acid. Before commencing testwork it is wise to consider the possible effects of the cation e.g. Ca^{2+}, possible accumulation or precipitation of salts and the cost of pH regulation in practice.

(f) <u>Water Quality</u>. Most preliminary testwork will employ either distilled water or local tap water. Many concentrators will employ different and probably much less pure water e.g. mine drainage, river water, well water. These often contain soluble ions or even organic species that complicate the flotation process and may even

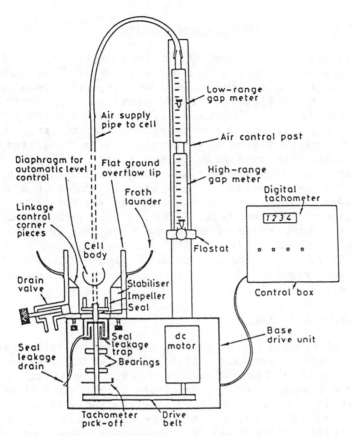

Figure 2. Leeds open-top cell.

vary seasonably (27,28).

(g) <u>Temperature</u>. Most testwork wil be conducted at room temperature but many processes in flotation, especially conditioning, are known to be affected by temperature. It is necessary to record pulp temperature and to consider the influence of temperature changes upon response together with information about the likely temperature changes at the plant site.

(h) <u>Pulp Density</u>. Many batch flotation machines have a cell volume of several litres but the weight of solids introduced is usually less than 1000 kg. Thus, most testwork is conducted at a pulp density less than that which economics dictate for plant practice i.e. 25 - 40 % by weight. In addition, the pulp density will be reducing significantly during the test. It is important to realise that the lower pulp densities are likely to result in cleaner separations i.e. higher concentrate grades with less entrained gangue.

4.3. Batch Flotation Tests – timed samples.

The timed collection of froth samples, provides the information from which kinetic data such as flotation rate constants can be calculated. In other respects, these tests are similar to the batch flotation tests previously described. The principal problem in using this kinetic data for the purpose of scale–up and plant design is the assumption that the rate constants are indeed constant when transferring from laboratory batch flotation machines to continuous plant scale machines. Factors such as particle size, bubble size, air/pulp mixing, aeration rate and froth structure amongst others are all known to affect the measured overall flotation rate constant and may not be the same in the lab cell and the plant cell. This has led to the use of "rules of thumb" for conversion of mean residence time in a lab cell to those in a continuous cell such as multiplication factors of between 1.5 for fast floating to 2 or even 3 for slow floating minerals. More rigorus mathematical procedures are considered in section (6.4).

4.4. Locked Cycle Tests.

A locked cycle test comprises a series of batch flotation tests so arranged to simulate a plant flowsheet probably containing stages of rougher flotation, scavenger flotation, cleaning flotation and even regrinding (see Figure 3). The important aspect of the test is that intermediate products and water are recycled according to the flow sheet until the weight and assays of the products attain constant values. Experience shows that this may take 6 or more complete "cycles". All the previous comments related to batch testing are relevant.

It is thus apparent that locked cycle tests require many man-hours of effort, extensive planning and organisation of the laboratory and rapid techniques of chemical analysis. Locked cycle tests should be seriously considered in the following circumstances.

(a) Previous batch test work has shown that a significant proportion of the value is contained in a middling product i.e. one of a grade unsuitable for marketing or disposal to waste. Obviously this product should be examined to determine whether the problem is one of liberation or misplacement but in either case further processing is ncessary. In many cases, sufficient comparable experience is available to indicate what fractions of the value in the middlings will eventually report to the concentrate and tailings, but where this information is unavailable a locked cycle test is necessary.

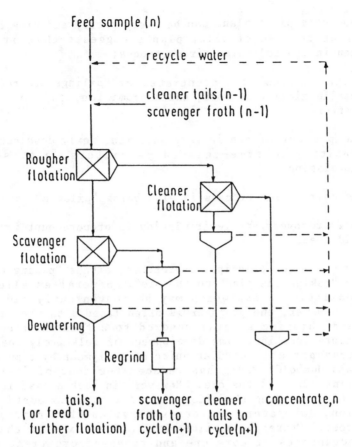

Figure 3.   Schematic locked cycle test, cycle n.

(b)   It is necessary to recycle as much process water as possible
and it is anticipated that this will incur process problems
e.g. by accumulation of reagents, soluble ions, degradation
products of reagents etc.

The principal disadvantages of this procedure are the time and
cost. The fact that during the test, which necessarily takes much
longer than the equivalent plant process, alteration of minerals
and reagents may occur can also cast doubts upon the results.

4.5.   Pilot Plant Tests.

A pilot plant can provide excellent information upon aspects
that are difficult to evaluate by other means e.g. scale-up, water
recovery and recirculation, intermediate products (middlings)
recirculation. The concensus of opinion seems to be that the
capacity should be at least 1 tph. However, the construction and

operation of a pilot plant can be very costly and time consuming. A review of the use of pilot plants suggests that it is only undertaken in the following circumstances;

(a)   large samples of concentrates or tailings are required for metallurgical test work, sales promotion, tailings disposal testwork etc.

(b)   the process or ore is very uncommon, newly developed or very sensitive to time-related parameters, e.g. selective floculation

(c)   the comminution process is also being "piloted"

(d)   water recovery and recirculation is of paramount importance - arid areas.

There are two principal approaches; either "taking the ore to plant" or "taking the plant to the ore". Several excellent pilot plant facilities exist which may be conveniently close to the deposit. However, the proposer is often faced with the problem of mining and shipping several hundred to a thousand tonnes of ore from a remote location. The development of relatively inexpensive, highly transportable modular units of equipment by companies such as Sala and Humboldt-Wedag has reduced the cost of installing a pilot plant close to the ore. However, in such a case it is often the cost of supporting the pilot plant, i.e. accomodation, power generation, laboratory and other services that represents the greater cost. Operating a pilot plant at the site of the deposit offers advantages of more ore and types of ore treated, use of local water, testing the reliability of equipment and involvement and training of local operators all of which may be important considerations.

The most important criticisms of pilot plant operation are as follows. It is usually manned and supervised by highly trained, highly motivated personnel and may achieve better results than can be achieved in commercial operation. It requires relatively large tonnages of ore for which drill core is generally insufficient such that it is only possible to treat near-surface or even outcrop ore or material from a limited number of exploration adits. It is extremely important to thoroughly explore the orebody by drilling or other techniques otherwise the pilot plant may process material largely unrepresentative of the orebody to be mined.

5.   STATISTICAL ASPECTS OF FLOTATION TESTWORK

In an ideal world each new ore would be subjected to the 5 types of test procedure in turn to gather the maximum amount of data for plant design and cost purposes. However, in practice, skilled use of types 4.1-4.3 can provide sufficient information for ores which respond similarly to existing industrial processes.

In simple situations batch testwork may be sufficient. The traditional approach is to test variables likely to influence response one at a time, striving to keep other parameters constant. For instance, a pH value may be chosen from experience and recovery of valuable mineral monitored as collector dosage is altered. If this shows some maximum or optimum response then pH is altered whilst keeping the collector at the constant level determined in the first set of tests. In this way an `optimum´ set of conditions can slowly emerge. This type of procedure can be cost effective but suffers from a number of drawbacks. Often it fails to indicate both the significance of changes relative to the experimental error and the presence of interactive effects between variables. For instance, pH changes may alter recovery by influencing the collector solubility or ionisation and the bubble size distribution in the cell at the same time. The correct application of statistical techniques can help a great deal in improving the validity of batch laboratory testwork as will be discussed below.

A major objection to batch testwork as traditionally carried out is that it supplies little kinetic data, even if a statistical design is used. The concept of rate is of great help in scale-up and design. In kinetic batch tests the froth is removed continuously but is analysed in separate timed batches so that the change in flotation response as the test proceeds can be monitored. This method can have distinct advantages as demonstrated by figure 4. Here two sets of recovery (R) values measured under different conditions for the same ore (say two different pH values) are plotted against time. It can be seen that although 2 floats much faster than 1 the final maximum achievable recovery is higher in case 1. This can have important implications for plant design and economic assessment and may have been missed had only the one-variable at a time batch testwork been carried out as explained below.

If the tests had been carried out using a single batch flotation time less than t_E then pH (2) would have given superior results whereas, if a time greater than t_E had be chosen, pH (1) would have been marginally better. Klimpel(29) has given a good discussion of this technique including a mathematical treatment of the results.

5.1. Planning the Experiment

Whatever type of experimental procedure is chosen the wise investigator plans his work to maximise the amount of information generated by each test and to provide checks on the validity of the conclusions. This does not mean that intelligent judgement should be suppressed but that it should augment a planned statistical

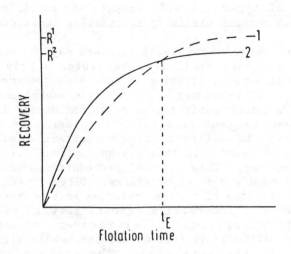

Figure 4. Recovery vs. time for batch flotation tests

design. Such a design, if properly planned and conducted, can make the most efficient use of the time and money available for experimental testwork.

A number of statistical procedures are useful in designing experimentation and these will be dealt with below. In this discussion the _response_ or dependent variable is the variable under investigation, which we usually hope to optimise. A _factor_ or independent variable is an experimental parameter which is varied by the investigator to determine its effect on the response. A factor can be qualitative (e.g. collector manufacturer) or quantitative (e.g. dosage). The value which a factor takes in any particular test is its _level_. A specific combination of factor levels is called a _treatment_ and a _test_ is the operation of the experiment at one treatment combination. _Precision_ is a measure of the spread of a measurement and poor precision implies uncertainty and reduced confidence in the result. _Accuracy_ is a measure of the deviation from the true result and inaccurate data implies _bias_, i.e. systematic error. _Random error_ is unpredictable and tends to a zero mean in the long run. _Randomising_ testwork means carrying out the tests in a programme in random order in an attempt to minimise the influence of unknown factors.

5.2. Significance Testing

Significance testing is a procedure for comparing quantities in a formal way designed to judge the significance of the observed differences in the quantities. It is based on formulating hypotheses and testing their validity. The null hypothesis (H_0) is that there is no significant difference between the quantities. The alternative hypothesis (H_1) is that there is such a difference. In general H_0 is assumed true unless proved otherwise.

5.2.1. The t-distribution

The t-distribution is an appropriate technique for small samples (n < 40) and is given by

$$t = (\bar{x} - \mu)(\sqrt{n})/S \qquad (1)$$

which $\bar{x}$ and S are the mean and standard deviation respectively.

This distribution has (n − 1) degrees of freedom and gives <u>confidence limits</u> of $\pm$ t.S/$\sqrt{n}$. The confidence limits are the two values between which the true result (μ) lies at a required level of certainty.

(a) <u>Method 1</u>.   x = 25.17 %

S =   0.32

n =   10

(b) <u>Method 2</u>.   X = 25.31 %

S =   0.28

n =   11

Table 1.   Sulphur Analysis of Feed Material

Table 1(a) shows the mean and standard deviation of 10 separate sulphur assays made on samples cut from the feed material used for a series of flotation tests. Using the value from standard tables for t with 9 degrees of freedom we can show that at the 95 % confidence level the real value of sulphur in the feed material lies between

$$\bar{x} \pm (t_{95,9})(S)/ \sqrt{n} \qquad (2)$$

that is μ lies between 25.17 $\pm$ (2.26)(0.32)/$\sqrt{9}$ or between 24.93 % and 25.41 %.

To test the significance of the difference between the means of two samples of size n_1 and n_2 having variances of S_1^2 and S_2^2, the t statistic is used but incorporating a pooled variance :

$$t = (\bar{x}_1 - \bar{x}_2)/S(\frac{1}{n_1} + \frac{1}{n_2})^{\frac{1}{2}} \qquad (3)$$

where $\quad S^2 = \dfrac{(n_1 - 1)\ S_1^2 + (n_2 - 1)\ S_2^2}{n_1 + n_2 - 2}$

The value of t is taken from the standard table with $(n_1 + n_2 - 2)$ degrees of freedom at the desired probability level.

Sample No	Recovery (%) Collector A	Collector B	Difference
1	84.2	86.3	−2.1
2	71.9	75.6	−3.7
3	77.8	76.6	1.2
4	80.1	82.6	−2.5
5	75.6	77.8	−2.2
6	70.8	70.0	0.8
7	81.3	85.2	−3.9
8	82.6	80.6	2.0
9	78.5	77.1	1.4
10	79.2	81.7	−2.5
Mean	78.2	79.3	−1.15

Table 2. Recovery with 10 samples, paired results

For example, table 1(b) shows the results of sulphur assays on a further 11 samples of the feed material but using a different analytical technqiue. Do the two methods give the same result? Here the null hypothesis is $H_o : \bar{x}_1 = \bar{x}_2$. The calculated value of t is

$$t = \frac{(25.27 - 25.31)}{(0.30)\left(\frac{1}{10} + \frac{1}{11}\right)^{\frac{1}{2}}} = -1.07$$

The standard value of t at the 95 % significance level is $t_{95,19} = 2.09$. Since the calculated t value is less than the standard value we conclude that there is no difference in the mean values obtained by either method. Note that the t distribution is symmetrical and the minus sign has no significance in the test.

Another useful variation of the t-test is for paired comparisons. Here the null hypothesis is that the differences between the pairs of results will be a sample from a population with mean zero and

$$t = \bar{d}/S \sqrt{n} \text{ with (n-1) degrees of freedom} \qquad (4)$$

where $\quad S^2 = \dfrac{\sum (d_i - \bar{d})^2}{n-1}$

and $\quad d_i = x_{i,1} - x_{i,2}$

Consider the case where two collectors A and B are investigated on 10 samples of ore from different parts of the orebody. Each sample is split in two and the sub-samples treated with each collector. The response variable is recovery and all other factors are kept constant. The results are given in table 2 and we wish to decide if there is any difference in the efficiency of the two collectors.

The differences have a mean of -1.15 with a variance of 5.052. The calculated value of t is

$$t = \frac{-1.15 \ (\sqrt{10})}{(2.25)} = -1.62$$

The standard value of $t_{95,5}$ is 2.26 and hence we have no evidence that there is any difference between the two collectors. Note that it would be wrong to compare the two mean recoveries (78.2 % and 79.3 %) directly since variations between ore samples will swamp any difference there may have been between the two collectors.

5.2.2. The F-test

The F-test is used to compare two variances i.e. compare the precision of two sets of data. $F = S_1^2/S_2^2$ where S_1^2 is always the larger variance. The value of F is compared to standard tables of the F distribution with $(n_1 - 1)$ and $(n_2 - 1)$ degrees of freedom. The null hypothesis is that there is no significant difference between variances with the alternate hypothesis that S_1^2 is greater than S_2^2.

Table 3 shows the feed rates of a collector solution into a pilot plant measured at intervals over an 8 hour shift using two different flow controllers. The target value was 125 cc/min to

Measurement	Flowmeter 1	Flowmeter 2
1	126	129
2	138	120
3	121	126
4	122	128
5	126	126
6	129	127
7	118	125
8	122	125
Mean	125.3	125.7
Variance	38.5	7.4

Table 3. Flowrate of Collector (cc/min)

which both controllers approximated on average. However, controller 2 apparently gave more consistent control having the lower variance. Since a consistent collector feed rate was important was the difference real or not? To find out the F value is calculated

$$F = 38.5/7.4 = 5.2$$

The standard F value with 9,9 degrees of freedom is $F_{95,9,9} = 3.2$. Thus the calculated F value is significant at the 95 % level and we reject the null hypothesis. Controller 2 is a more consistent feeder.

In all the tests described above we have used the 95 % confidence level which means we have 95 out of 100 chances that our conclusion will be correct. We could, of course, draw the wrong conclusion 5 % of the time. In some instances engineers may wish to be more certain that the risk of making a wrong decision is small and test at the 99 % level. There is also a distinction between what is statistically significant and what is practically important. The difference in the precision of the two flowmeters is significant at the 95 % level but whether the engineers regard the actual fluctuations as important is a matter of professional judgement.

5.3. Analysis of Variance (ANOVA)

The ANOVA technique is used to determine whether several means are significantly different from one another when compared to the experimental error. It uses the F-test to assess the variance due to changes in factor level. Either one or two factors can be varied giving rise to one-way or two-way ANOVA. The data layouts and calculation procedures are summarised in tables 4 and 5. For detailed descriptions of the derivations of the expressions used the reader is referred to one of the many texts on the subject (30,31).

In general, the between-sample sum of squares which results from real differences between samples is calculated and compared to the with-in sample sum of squares, this being due to experimental error. To determine the significance the two sum of squares are compared in an F-test after allowing for the degrees of freedom attached to each.

5.3.1. Flotation test. One-way ANOVA

As an example, consider the results of a series of 15 flotation tests carried out on an ore and given in table 6. Here the response variable is zinc recovery and 3 repetitions are carried out at five collector levels. The question asked is "Does an increase in collector addition lead to a genuine increase in zinc recovery?"

Collector Conc (g/t)	Zinc Recovery (%)		
	1	2	3
50	81.5	79.6	81.4
75	82.3	81.6	83.1
100	83.4	83.1	82.9
125	85.6	85.2	84.1
150	85.1	84.2	84.6

Table 6. Flotation Tests. One-way ANOVA

ANOVA table

Source of Variation	Sum of Squares	Degrees of Freedom	Mean Square	F
Between collector levels	34.59	4	8.65	16.8
Within collector repetitions (experimental error)	5.15	10	0.51	

Data Layout

Factor Level	Repetitions	Total	Observations*
1	$x_{11}, x_{12} \ldots \ldots x_{1n_1}$	S_1	n_1
2	$x_{21}, x_{22} \ldots \ldots x_{2n_2}$	S_2	n_2
k	$x_{k1}, \ldots \ldots \ldots \ldots x_{kn_k}$	S_k	n_k
		S	N

* not necessarily equal for each factor level.

Calculation

Source of Variation	Sum of Squares	d.f.	Mean Square	F
Between levels	$S_F = \sum \left(\frac{S_i^2}{n_i}\right) - C$	k-1	$M_F = \frac{S_F}{k-1}$	M_F/M_E
Within levels	$S_E = \sum (x_{ij}^2) - \sum \left(\frac{S_i^2}{n_i}\right)$	N-k	$M_E = \frac{S_E}{N-k}$	
Total	$S_T = \sum (x_{ij}^2) - C$	N-1	-	-

where $C = S^2/N$

Table 4. One-way ANOVA. Data Layout and Calculation

Data Layout

		Factor A			Row Total
	Level	1	2	P	
Factor B	1	$x_{111} - x_{11r}$	$x_{121} - x_{12r}$	$x_{1p1} - x_{1pr}$	R_1
	2				R_2
	q	$x_{q111} - x_{q1r}$			R_q
Column Total		C_1	C_2	C_p	T

Calculation

Source of Variation	Sum of Squares	d.f.	Mean Square	F
Main effect A	$S_A = \sum \frac{C_j^2}{rq} - C$	p-1	$M_A = \frac{S_A}{p-1}$	M_A/M_R
Main effect B	$S_B = \sum \frac{R_j^2}{rp} - C$	q-1	$M_B = \frac{S_B}{q-1}$	M_B/M_R
Residual	$S_R = \sum (x_{ijk})^2 - \sum \frac{T_{ij}^2}{r}$	pq(r-1)	$M_R = \frac{S_R}{pq(r-1)}$	
Interaction	S_I (By subtraction)	(p-1)(q-1)	$M_I = \frac{S_I}{(p-1)(q-1)}$	M_I/M_R
Total	$S_T = \sum (x_{ijk})^2 - C$	pqr-1		

when $C = T^2/pqr$ and T_{ij} is the sum of observations in the (i,j)th cell.

Table 5. Two-way ANOVA. Data Layout and Calculation

The standard value of $F_{4,10}$ is 3.48. Hence our calculated value exceeds this and the increase in recovery obtained with increasing collector concentration is real and significant.

5.3.2. Flotation Test 2-way ANOVA

Two factors can be varied in this technique and the influence of factor A, factor B and any interaction (AB) can be assessed by comparison with the experimental error (residual). The following simple example will illustrate the procedure.

Factor B. pH	Factor A Collector Addition (g/t)	
	50	100
10	(83.2)(85.9)	(82.0)(87.3)
11	(94.1)(91.8)	(88.3)(90.0)

Table 7. Copper Recovery (%). Analysis by 2-way ANOVA

Table 7 gives the results of 4 flotation tests carried out with two replicates and at two levels of collector concentration and pH. The response variable is recovery of copper into the rougher concentrate. Following the data layout given in table 5 the following ANOVA table is calculated:

ANOVA table

Source of Variation	Sum of Squares	Degrees of Freedom	Mean Square	F
Main Effect (A)	6.85	1	6.85	1.26
Main Effect (B)	83.21	1	83.21	15.28
Interaction AB	7.60	1	7.60	1.40
Residual (error)	21.78	4	5.45	–
Total	119.44	7		

The calculated F values are compared to the standard value of $F_{95,1,4} = 7.71$ and show that only main effect B (pH) has a significant influence over the range tested. Note that the residual component is due to experimental error as determined by the replicates and would not exist if the experiment contained no replication. Of course, the example given was kept brief for clarity and the ranges of the two factors tested could have been much greater. This type of analysis allows an economy of effort in that the effect of two factors is investigated simultaneously in a structured statistical design. Moreover, interactions, if they exist, could be recognised and investigated as could the reasons for an overlarge residual mean square, indicative of poor experimental technique.

5.4. Reduction of factors

If the investigator were to list (as indeed he should) those factors that might influence the outcome of an individual flotation test he would come up with a very long list indeed. Experience and a few preliminary tests may enable him to confidently reduce the number somewhat but testing those left would still represent a formidable task. One way of statistically deciding whether or not a factor has a significant effect on the outcome is to use a factorial design analysed by Yates technique. In this procedure each of n factors is tested at two levels, a 'high' value and a 'low' value (which may be zero) covering a sensible range relative to the definition of the problem. Such a design gives rise to 2^n experiments and is called a 2^n factorial design. The influence of each factor is tested by an F-test with the calculated F values being derived by an algorithm due to Yates (31). The calculations are fairly simple and the results allow a preliminary screening to be made of those factors which have a significant effect and should be investigated more thoroughly. At first sight the number of tests involved appears to be large, for instance when 8 factors are desired, $2^8 = 256$ experiments are indicated. To circumvent this problem fractions (subsets) of the full design can be used. There will be a loss of information but this can be selected to relate to interactions rather than main effects and the overall work load greatly reduced with little reduction in the basic information obtained.

5.5. Calculation of results, criterion of success and error.

In addition to the significance of the factors tested relative to experimental error there is often a need to judge the success of an experiment or set of experiments. The criterion of success may be a simple maximisation of recovery or grade of one component. More complex definitions such as enrichment ratio are also used (32) with one of the more useful definitions being that of separation efficiency, E_s, where E_s is given by

$$E_s = (R - R_g)$$

Here, R is the recovery of valued constituent and R_g the recovery of gangue

Results are often assessed by economic performance and this will involve the calculation of some success parameter utilising the test results and other quantitative data such as smelter costs, freight rates etc. It must be recognised that experimentally derived values used in such calculations will induce errors in the calculated success parameter. These errors are carried through the calculation by what is termed propagation of error, the consequences of which can sometimes be unexpected.

It can be shown that if some value z is calculated using a function containing the experimentally determined values w, x and y then, to the first approximation

$$\delta z^2 = \left|\frac{\partial f}{\partial w}\delta w\right|^2 + \left|\frac{\partial f}{\partial x}\delta x\right|^2 + \left|\frac{\partial f}{\partial y}\delta y\right|^2 \qquad (5)$$

where $\delta z, \delta w, \delta x, \delta y$ are the maximum errors in z, w, x and y. $\partial f/\partial x$ is the partial differential of the function $z = f(x,y)$ with respect to x.

For example, consider a pilot plant being operated on a barite ore and that the measure of success is the yield of concentrate (R). The experimentally obtained values are the assays of the feed (f), concentrate (c) and tailing (t) respectively. R is calculated using the two product formula

$$R = \frac{100c\ (f-t)}{f(c-t)} \qquad (6)$$

The values are f = 40.4 %, c = 84.1 % and t = 7.3 % and each value has a maximum error of $\pm$ 0.5 %. (The definition of `maximum error' is loose but it can be taken to mean plus or minus two standard deviations, $\pm 2S$). What is the maximum likely error in the calculated value of R? One may suppose that it would be ± 0.5 % or would be given by the combining variances through $S_R^2 = (S_f^2 + S_c^2 + S_t^2)$ but both would be wrong. In this problem $\delta f = \delta c = \delta t = 0.5$

Also $\dfrac{\partial R}{\partial c} = \dfrac{100\ t\ (f-t)}{f(c-t)^2} = 0.101$

$\dfrac{\partial R}{\partial t} = \dfrac{100\ c\ (c-f)}{f(c-t)^2} = 1.542$

$\dfrac{\partial R}{\partial f} = \dfrac{100\ c\ t}{f^2\ (c-t)} = 0.490$

and R = 89.7 %.

Using equation (5) we can calculate

$$\delta_{R^2} = (0.490)^2(0.5)^2 + (0.101)^2(0.5)^2 + (1.542)^2(0.5)^2 = 0.811 \%$$

In other words the calculated value of recovery (R = 89.7 %) has a maximum error of ± 0.81 %, higher than the individual values from which it was calculated. It can also be seen that three components of the total error can be examined separately to determine which contributes the most. In this case improvement in the precision of R is best achieved by improving the error of the measured value of t.

The experimenter should also note that the error in R depends both on the errors in the measured values <u>and</u> the magnitude of these values. Consider the case where the pilot plant was run with a set of conditions that gave poor recovery, say with f = 40.4 %, c = 86.7 % and t = 32.1 %, each of these values also having a maximum error of $\pm$ 0.5 %. Calculating the relevant parameters we have

$$R = 32.6 \%; \frac{\delta R}{\delta c} = 0.221; \frac{\delta R}{\delta t} = 3.930; \frac{\delta R}{\delta f} = 3.123$$

thus giving $\delta R = (5.025)(0.5) = 2.5 \%$

Thus, the error has deteriorated both in its absolute value and in terms of relative error ($\delta R/R$) despite the measured values having the same precision.

Calculations of simple parameters like recovery are often carried out during the course of a flotation investigation but rarely is the error of the calculated value assessed. Obviously this may have repercussions in many situations but especially when optimisation is being attempted and real increases in recovery, say of 1 %, are deemed to be important.

6. MODEL DEVELOPMENT AND OPTIMISATION

6.1. Kinetic Models

6.1.1. Batch tests

Research has shown that flotation can generally be regarded as a first-order rate process (33) with respect to the concentration of the floating species in the pulp, c_i. Hence, a batch flotation experiment will have

$$dc_i/dt = - kc_i$$

$$\text{or } c_i = c_o e^{-k_i t} \tag{7}$$

where c_o is the initial (feed) concentration

k_i is the first order rate constant

t is the time elapsed since the start of flotation.

After some period of flotation, t, it can be seen that recovery of species i (R_i) is given by

$$R_i = \frac{c_o - c_i}{c_o} = 1 - e^{-k_i t} \qquad (8)$$

The flotation rate constant can then be found by plotting ln (1 − R_i) against t and reading off the slope of the best-fit straight line. Obviously, one hopes to arrange, by suitable selection of reagent additions, air flow rate etc. that k values for gangue minerals are much lower than those for a valuable floating species. The range of values usually found for k is between 0.01 and 1 min^{-1}.

Of course, some gangue will float and dilute the concentrate. The longer flotation continues the more will be this dilution and cumulative concentrate grade will decrease. To maximise the difference in recovery between two minerals we wish that $R_1 - R_2$ is a maximum. Hence, if E = $R_1 - R_2$ = [1 − exp (−$k_1 t$)] − [1 − exp (−$k_2 t$)]

we can show that when dE/dt = 0

$$k_1 \exp(-k_1 t) = k_2 \exp(-k_2 t)$$

from which $$t = \frac{\ln k_1 - \ln k_2}{(k_1 - k_2)} \qquad (9)$$

It is evident that there is an optimum flotation time beyond which the incremental increase in recovery of (2) begins to exceed that of (1). Table 8 shows what happens when k_1 = 0.3 and k_2 = 0.01 min^{-1}. Equation (9) predicts an optimum flotation time of 11.7 min and it can be seen that after this time E starts to decrease. Of course, recoveries of 99 % in practice are rare with most minerals showing a maximum recovery, R_m, after `long´ but `reasonable´ flotation times. Equation (8) can be modified to take account of this by

$$R_i / R_{m,i} = 1 - e^{-k_i t}. \qquad (10)$$

As a general approach, the above is a good approximation of what happens with a simple two component ore where species 1 is the valuable mineral and species 2 the gangue. Of course, most ores have several waste minerals that have significant (and different) values of k_2. In this case, k_2 can be taken as a weighted average of the individual rate constants. That is, there are two single values of k, for the valuable species and for `the rest´. This approach can introduce errors due to mineralogical differences between test samples.

$$k_1 = 0.3 \ min^{-1}$$

$$k_2 = 0.01 \ min^{-1}$$

t(min)	R_1	R_2	E
1	0.26	0.01	0.25
4	0.70	0.04	0.66
7	0.88	0.07	0.81
10	0.95	0.10	0.86
13	0.98	0.12	0.86
15	0.99	0.14	0.85

Table 8. Optimum Flotation Time

To ensure that differences between samples are not hiding differences in response due to factor level changes is difficult. However, analysis can be aided by a regression approach. The observed recovery values over time are modelled by

$$\ln \frac{(R_{m,i} - R_i)}{R_{m,i}} = k_i t \tag{11}$$

using a linear regression, least squares criterion. The best fit values of R_m and k_i are determined with their respective confidence levels. It is thus easier to determine whether or not real differences in k_i or R_m values exist as the treatment changes.

In an attempt to avoid the problem of individual rate constants for several floating species, some workers have used models of the type

$$R_i/R_{m,i} = 1 - \left\{ \frac{1 - \exp(-kt)}{kt} \right\} \tag{12}$$

where k is the first order rate constant of total mass removal (value plus gangue).

R_i, $R_{m,i}$ are the recoveries of valuable species. Again best values of $R_{m,i}$ and k may be found by curve-fitting. In these models k has a direct physical meaning in that it represents the rate of mass removal from the cell and may be used for design purposes.

A few words of caution are necessary at this point. Models of batch flotation of the type described have wide application but some ores may require more complex forms. This is particularlly true if relatively large amounts of coarse or fine floating material exist in the feed since k_i is known to vary with particle size. In these cases particle size fractions may have to be modelled separately.

A method of prediction and simulation of locked cycle tests has been proposed that uses a simplified concept (34). Here `split factors´ are asigned to every component in each separator and the circuit simulated by using an algorithm. A split factor is the fraction of each mineral species in the feed to each separator that reports to the tailings. Batch-generated split factors are used to initiate the procedure which is brought to a steady state as defined by some convergence criterion. Surprisingly, for many ores, the assumption that split factors remain constant as material proceeds from stage-to-stage appears to remain valid.

6.1.2. Pilot Plant Kinetic Model

A convenient method of modelling a continuous pilot plant cell is to use the population balance model technique (35). In this method the fluid flowing into the cell is considered to comprise a large number of identical fluid `elements´, each containing a perfect sample of the feed slurry. These elements pass through the cell and floatable material is removed. The total amount of floatable material removed depends on how long each fluid element stays in the cell; its residence time. The residence time of fluid elements depends on the hydrodynamics of the system and can be represented by a frequency distribution of residence times, $E(t)$. $E(t)$ can take many forms but for most flotation cells it approximates to perfect mixing (36) where

$$E(t) = \frac{e^{-t/\bar{t}}}{\bar{t}} \tag{13}$$

and $\bar{t}$ is the fluid mean residence time.

The final concentration of floatable material leaving a cell is hence an integration of 1st order removal of floatable material from elements with the residence times existing in the system. In symbols, $c = \int_0^\infty c_E . E(t)dt$ where $c_E = c_o e^{-kt}$ and t is the residence time of an element

$$\text{Hence} \quad c = \int_0^\infty \frac{(c_o e^{-kt})(e^{-t/\bar{t}})}{\bar{t}} dt \tag{14}$$

$$\frac{c}{c_o} = \frac{1}{1 + k\bar{t}} \tag{15}$$

Recovery is then $R = 1 - c/c_o = \dfrac{k\bar{t}}{1 + k\bar{t}}$ (16)

Thus, if the steady state recovery of a pilot plant cell is monitored and t is known then k values can be calculated from (16). In many cases the variation in k for different particle sizes of a given mineral requires that each size is monitored and modelled indpendently although to a first approximation the concept of two classes, `slow floaters´ and `fast floaters´ may be adequate (37). If a valuable component occurs in more than one phase a complex distributed parameter model of k may be necessary (38), for instance when a mineral is not completely liberated.

6.2. Factor Models

The kinetic models described above are most useful but often simpler steady state relationships can suffice. In these the factors are held at fixed settings which do not fluctuate, so called non-stochastic or deterministic models.

When factors and responses are continuous in scale they may be empirically modelled by simple first or second degree polynomials. For instance, in the case of a single factor, say collector concentration (X) the recovery (Y) obtained in a series of tests may follow

$$Y = a_o + a_1X \qquad \text{first degree} \qquad (17)$$

$$\text{or} \quad Y = a_o + a_1X + a_2X^2 \qquad \text{second degree} \qquad (18)$$

Figure (5) shows these two cases and the values of the model parameters (a_o, a_1, a_2) may be found by a least-squares regression technique. Note that for a second degree polynomial the maximum value of Y occurs at an X value of $2a_2/a_1$ and up to this value a first degree model may be sufficient.

If two or more factors are modelled then account may have to be taken of any interactions. If no interaction exists the factors are said to be additive, that is the effect of a single variable at selected fixed conditions of other variables is the same as at other settings of the other variables. Figure (6) shows an additive (non-interactive) response to collector (X_1) and pH (X_2). A suitable model is

$$Y = a_o + a_1X_1 + a_2X_2 \qquad (19)$$

However, figure (7) shows a case where considerable interaction has occurred and the model will contain an X_1X_2 term, thus,

$$Y = a_o + a_1X_1 + a_2X_2 + a_3X_1X_2 \qquad (20)$$

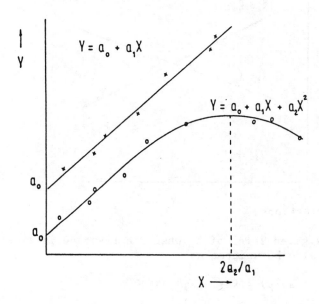

Figure 5. Polynomial models.

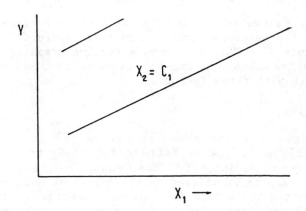

Figure 6. Additive models.

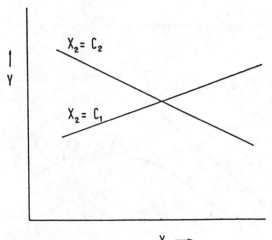

Figure 7.   Interactions.

Note that for a given level of X_2 when the response is linear, this expression can also be written as

$$Y = (a_0 + a_2X_2) + (a_1 + a_3X_2)X_1 \tag{21}$$

where $(a_0 + a_2X_2)$ = intercept and $(a_1 + a_3X_2)$ = slope.

When the graphical relationship between interactive factors gives curves rather than straight lines the full second degree polynomial must be used.

$$Y = a_0 + a_1X_1 + a_2X_2 + a_3X_1^2 + a_4X_2^2 + a_5X_1X_2 \tag{22}$$

A range of values of X_1 and X_2 may best be represented as a contour map of the response surface Y over the space defined by X_1 and X_2 (see figure X). This is a very effective graphical overview of experimental results and can be extended to 3 dimensional surfaces for work with three factors.

6.3. Optimisation

The response surface shown in figure (8) is a method of finding the optimum levels of factors X_1 and X_2 in terms of the value of the response variable Y. The `shape` is that of a hill and optimisation may be visualised as the act of ascending the hill to find the summit. The method is somewhat laborious because of the large number of tests involved to obtain a clear idea of the position of contours. Also the resolution may be poor in the presence of experimental error and certainty about the site of the summit is often lacking. An experimental design involving a 2-way ANOVA will give information on the experimental error variance and can help interpretation.

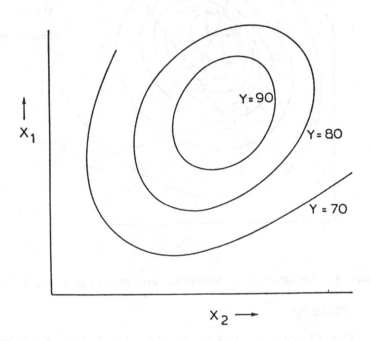

Figure 8. Contour mapping

To reduce the amount of work involved a 2-level factorial design may be used with the strategy of <u>steepest ascent</u>. A region is examined by a 2^2 factorial design and analysed (by Yates technique for instance) to show the direction of a positive increase in the response variable (Y). Another 2^2 design is repeated in a new region along the predicted direction until a decrease is observed. A new direction of `positivity´ in Y will be indicated and followed and so on. The procedure is illustrated in figure (9). Eventually a point should be reached where the influence of the two factors is small. This is evidence that the optimum region is close and one may now carry out a more detailed ANOVA over this small region to determine the optimum point.

This procedure can be very successful but often fails because of changes in Y induced by unknown (and uncontrolled) factors swamping the true direction indicated by response to X_1 and X_2. The engineer must have tight control over independent variables, hold all except X_1 and X_2 constant and avoid bias in his experimental technique and analytical systems.

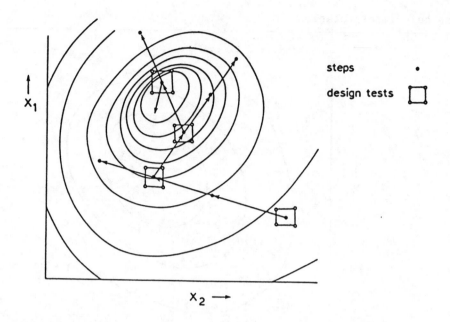

steps •

design tests ☐

Figure 9. Technique of steepest ascent using experimental designs.

6.4. Scale-up

The flotation engineer is faced with scale-up problems whenever he proceeds from small scale to large scale equipment. A number of dimensionless relationships are available to allow intelligent scaling of cell parameters such as airflowrate and impeller motor rating (39). However, a real problem exists when attempts are made to scale-up from batch laboratory work to continuous pilot or full scale equipment. This will be illustrated by using the simple first order flotation model discussed above.

Assume that batch flotation testwork has been carried out on an ore and the mass proportion recovery (R_B) of a floatable species is related to the time of flotation in a batch laboratory cell by

$$R_B = 1 - e^{-kt} \tag{23}$$

Here k is a lumped 1st order flotation rate constant whose value is found by plotting $\ln(1 - R_B)$ vs t for various flotation times. Consider now achieving the same recovery in a single continuous cell being fed with the same material. How large should it be? A convenient measure of size is the mean residence time of the cell $\bar{t}$ given by $\bar{t} = V/Q$ where V is the internal volume and Q the volumetric flowrate though it. Most flotation cells approximate to perfect mixers and it has been shown that the recovery of such a cell R_c is given by $R_c = k\bar{t}/(1 + k\bar{t})$

Equating R_B and R_c we get

$$1 - e^{-kt} = k\bar{t}/(1 + k\bar{t})$$

or

$$\frac{(e^{kt} - 1)}{k} = \bar{t}$$

The ratio $\bar{t}/t$ is then equal to $(e^{kt} - 1)/kt$ (24)

Batch Recovery In Time t (R_B)	kt*	$\bar{t}/t$
0.85	1.90	3.0
0.90	2.30	3.9
0.92	2.53	4.6
0.95	3.00	6.3

* from equation (24)

Table 9. Ratio of flotation time. Continuous to batch recovery

Table 9 shows the values of $\bar{t}/t$ for various values of R_B and it can be seen that the continuous flotation time $\bar{t}$ is several times that required in a laboratory batch cell. If 8 minutes were required in the batch test to get 92 % recovery, a time of (8)(4.6) = 37 mins would be required in a single continuous cell. If the volumetric feedrate Q is fixed then the cell volume can be determined through $V = Q\bar{t}$. It would be quite wrong to design the continuous cell with the same flotation time as that found in the laboratory, even with a reasonable safety factor. Note also that table 8 shows that the ratio $\bar{t}/t$ gets bigger the higher is R_B so the error in simply taking the laboratory flotation time gets even worse.

This increase in flotation time for continuous cells is associated with the loss of efficiency due to the distribution of residence times in perfect mixing. It can be compensated for to a great extent by using several cells in series to give the same overall residence time.

It can be shown (40) that if n individual flotation machine were arranged in series, each machine being identical, then as long as k remains constant the overall recovery R_c^n is given by

$$R_c^n = 1 - \frac{n^n}{(n + k\bar{t}c)^n} \qquad (25)$$

where $\bar{t}_c$ is the <u>total</u> mean residence time of a bank of n cells. If we now put $R_B = R_c^n$ then

$$\bar{t}_c/t = \frac{n}{kt} \left[\exp \left\{ \frac{kt}{n} \right\} - 1 \right] \qquad (26)$$

For example, if the batch recovery is 90 % then $R_B = 0.9$ and kt = 2.30. The ratio t_c/t can be calculated for various numbers of cells in series (n).

No. of machines (n)	$\bar{t}_c/t$
1	3.9
2	1.9
3	1.5
5	1.3
7	1.2

Table 10. Continuous recovery in n-machines $(R_B = R_c^n = 0.90)$

Table 10 shows that there is a substantial drop in overall retention time to achieve the recovery as n increases. Hence, if it required 6 minutes to achieve 90 % recovery in the laboratory batch cell then 6(1.5) = 9 min total flotation time would be needed using 3 cells in the equivalent continuous plant. Each cell would have a mean residence time of 3 mins and its volume could be determined for a given value of Q. It can also be seen that more than five cells in series produces little reduction in total residence time.

References

1. MacDonald R, Brison R, "Applied Research in Flotation" contained in "Froth Flotation" 50th Anniversary Volume. Ed.Fuerstenau D. (AIME, New York, 1962) p.298-323.

2.   Griffith W.A.   The design and analysis of flotation
      experiments, Froth Flotation 50th Anniversary Volume (New
      York, AIME, 1962) 277 - 297.

3.   Scobie A.G. and Wyslousil D.M.   Metallurigcal testing,
      design, construction and operation of the Lake Dufault
      treatment plant.   CIM Bulletin (April 1968) 482 - 488.

4.   Neumann G.W. and Schnarr I.R.   Concentrator operation at
      Brunswick Mining and Smelting Corporation No. 12 Mine, CIM
      Bulletin (Sept. 1971) 51- 61.

5.   Abu Rashid A.R. and Smith M.R.   Development of a selective
      flocculation - froth flotation process to beneficiate a
      non-magnetic iron ore of the Kingdom of Saudi Arabia, 14th
      IMPC (Toronto, CIM, 1982) Paper 10-1.

6.   Lewis P.J. and Martin G.J.   Mahd adh Dhahab gold-silver
      deposit, Saudi Arabia: Mineralogical studies associated with
      metallurgical process evaluation Trans. I.M.M. 92 (June 1983)
      C63-72.

7.   Jones M.P.   Practical Mineralogy (London, Graham and Trotman
      Ltd, 1984).

8.   Jones M.P.   Automatic Analysis, Physical methods in
      determinative mineralogy, 2nd Edition, Ed. Zussman J (London,
      Academic Press, 1977) 167 - 199.

9.   Jowett A., Heyes G.W., Risla G.   Investigation of lead
      recovery problems in the lead-zinc concentrator at Mount Isa,
      Queensland, Australia.   Trans. I.M.M. 90 (1981) C73 - C80.

10.   Sutherland K and Wark I Experimenal methods, Principles of
      flotation (Melbourne, A.I.M.M., 1955) 50 - 64.

11.   Hallimond A.F.   Mining magazine Vol. 70 (1944) 87 - 90 and
      Vol. 72 (1945) 201 - 206.

12.   Warren L.J., Kitchener J.A.   Role of fluoride in the
      flotation of feldspar : adsorption on quartz, corundum and
      potassium feldspar.   Trans. I.M.M. 81 (1972) C137 - C147.

13.   Iskra J., Gutierrez C., Kitchener J.A. Influence of quebracho
      on the flotation of fluorite, calcite, hematite and quartz
      with oleate as collector.   Trans. I.M.M. 82 (1973) C73 - 78.

14.   Page P.W.   Ph.D. Thesis, University of London (1984)

15.   Fuerstenau D.W., Pradip, L.A., and Raghavan S.   An
      alternative reagent scheme for the flotation of Mountain Pass

rare-earth ore 14th IMPC (Toronto, CIM, 1982) Paper 4-6.

16. Collins D.N. Investigation of collector systems for the flotation of cassiterite. Trans. I.M.M. 76 (1967) C77 - C93.

17. Coleman R.L., Metallurgical testing procedures, Mineral processing plant design, Ed. Mular A and Bhappu R (New York, AIME, 1980) 144 -181.

18. Pryor E.J., Mineral processing, 3rd Edition, (London, Elsevier, 1965) 622 - 626.

19. Armstrong D.G. Open and Closed circuit grinding on a laboratory scale. 5th IMPC (London, I.M.M., 1960) 67-78.

20. Trahar W.J. A rational interpretation of the role of particle size in flotation. Int. Journal of Mineral Processing 8(1981) 289-37.

21. Dell C.C., Release analsyis; A new tool for ore dressing research, Recent developments in mineral dressing (London, I.M.M., 1960) 75-84.

22. Dell C.C., Bunyard M.J., Rickleton R.A. and Young P.A., Release analysis; a comparison of techniques, Trans. I.M.M. Vol. 81 (1972) C89-C96.

23. Cutting, G.W., Barber, S.P., Watson, D. Prediction of plant performance from batch tests using process models: Effects of froth structure, 14th IMPC (Toronto, CIM, 1982) Paper 6-14.

24. Lynch A.J., Johnson N.W., Manlapig, E.V., Thorne, C.G., The pattern of behaviour of sulphide minerals and coal in flotation circuits, Mineral and coal flotation circuits (London, Elsevier, 1981) 21-43.

25. Dell C.C., and Bunyard and M.J. Development of an automatic flotation cell for the laboratory, Trans. I.M.M. 81 (1972) C246 - C248.

26. Dell C.C. and Hall G.A., Leeds open-top laboratory flotation cell, Trans I.M.M. 90 (1981) C174-C176

27. Paananen A.D. and Turcotte W.A. Factors influencing selective flocculation - desliming practice at Tilden Mine, Mining Engineering (Aug. 1980)1244 - 1247.

28. Alfano G, Carta M, Del Fa` C, Ghiani M, Rossi G, Recycling of residual waters in the flotation of non-sulphide minerals: experiences in the laboratory and continuous pilot plant, 14th IMPC (Toronto, CIM, 1982) Paper 5-8.

29.   Klimpel R.   Selection of chemical reagents for flotation.   In
       Mineral Processing Plant Design Eds. Mular A and Bhappu R.
       A.I.M.E., (New York) 1980, 907 - 934.

30.   Chatfield C.   Statistics for Technology (London, Chapman and
       Hall, 1978).

31.   Box G.   Statistics for Experimenters.   (New York, Wiley,
       1978).

32.   Schulz N.   Separation Efficiency.   Trans. A.I.M.E., $\underline{247}$
       (1970) 81 - 87.

33.   Huber-Panu I.   Mathematical models of batch and continuous
       flotation.   In Flotation, A.M. Gaudin Memorial Volume, Ed.
       Fuerstenau M.   (New York, A.I.M.E., 1976), p.679.

34.   Agar G and Kipkie W.   Predicting locked cycle flotation test
       results from batch data.   C.I.M. Bulletin (Nov. 1978) 119 -
       125.

35.   Himmelblau D. and Bischoff K.   Process Analysis and
       Simulation. (New York, Wiley, 1968) 59 - 83.

36.   Niemi A.   A study of dynamic and control properties of
       industrial flotation processes.   Acta Polytech Scand., $\underline{48}$
       (1966) 111.

37.   Lynch A.   Mineral and Coal Flotation Circuits (Amsterdam,
       Elsevier, 1981).

38.   King, R.   A pilot plant investigation of a flotation model.
       N.I.M.   Report No. 1573 (Johannesburg, 1974).

39.   Harris, C.   Flotation machines. In Flotation. A.M. Gaudin
       Memorial Volume.   Ed. Fuersteanau M. (New York, A.I.M.E.,
       1976) 753 - 815

40.   Jowett A and Sutherland D.   A simulation study of the effect
       of cell size on flotation costs.   The Chemical Engineer (Aug
       1979) 603 - 607.

MINERAL PROCESSING FLOWSHEET DEVELOPMENT

Jacques A. De Cuyper
Professor
Laboratoire de Traitement des Minerais
Université Catholique de Louvain
1348 Louvain-la-Neuve, Belgium.

ABSTRACT

Metallurgical testing is the most essential part of any mineral processing flowsheet development. Its objectives must therefore be clearly defined. These are discussed together with the importance of the choice of the samples required for the test program.

Rather than describing testing procedures, stress is laid on how the test program should be established and what informations such planning requires.

Recommendations are given concerning the choice of the process flowsheet on the base of the laboratory batch test results. Finally, while cycle tests should be considered as useful additional tests and be recommended, pilot plant testing is open to much controversy : its difficulties and advantages are presented.

1. INTRODUCTION

In most cases, the raw materials treated by mineral processing are natural ores, i.e. highly heterogeneous mixtures of solid mineral species. Their heterogeneity is such that not only the texture of the ore (and consequently the size of liberation), the type of minerals present and their relative amounts, but even the properties of similar minerals may not be constant throughout the same ore body.

Therefore, while it might be feasible, without any preliminary testwork, to develop a complete process flowsheet based on smelting operations as soon as the elemental analysis of the feed is known, on the contrary it is absolutely required to perform preliminary laboratory testing whenever the flowsheet involves mineral processing operations. In contrast to smelting, which has an effect of washing out the mineralogical features of the ore, mineral processing is indeed essentially dependent on the mode of occurrence of the valuable elements, whether it uses physical methods of separation or leaching.

The purpose of this paper is to discuss the philosophy of mineral processing flowsheet development and to make recommendations based on the experience accumulated since 25 years in the author's laboratory through metallurgical testwork performed in close relation with specialized engineering firms engaged in feasibility studies on new mineral deposits.

2. OBJECTIVES OF METALLURGICAL TESTING

The objectives should be clearly defined by the client. Of course, they may differ according to the stage of perfection to which the feasibility study has to be brought, either preliminary or final. The client may also have very good personal reasons for requesting to limit the testwork to some types of processes or to the obtention of well specified products. Nevertheless, even in these cases, orientation tests determining the metallurgical, chemical and physical properties of the ore should never be dropped out of the program. Such tests indeed permit to appraise the well-founded character of the client's request and to draw his attention upon interesting particularities he might not be aware of.

But the main objective of metallurgical testing is of course the establishing of the process flowsheet, with determination of the material balance, metallurgical recoveries and quality of the final products. The metallurgical testing should also provide all necessary information on the operating parameters for design purposes and capital and operating costs estimates.

By stating these objectives, it is clear that, when involved in the developing of a mineral processing flowsheet, the metallurgist should already search for the overall economics when choosing a process or combination of processes.

3. CHOICE OF THE SAMPLE(S) FOR METALLURGICAL TESTING

The reliability of the process flowsheet of course depends on the representative character of the sample on which the flowsheet was based. The sample collected for flowsheet development investigations should therefore represent the "average" ore of the deposit, with respect not only to chemical composition, but also to mineralogical composition and texture (i.e. grain shape and size distribution, nature of gangue-ore boundaries and degree of dissémination of the valuable minerals).

Metallurgists might, at first sight, not worry too much about the true representativeness of the sample, considering the fact that the exploration geologists involved in the ore reserve evaluation would evidently be well aware of the importance of the sampling. In fact, for the geologist, the sampling of a mineral deposit provides data for estimating "ore reserves" including grade and tonnage. But the metallurgist's concern is the feed to the mill. The sample he needs should thus correspond to the "mining reserves", which are based on cut-off grade and take into account the actual mining method with its eventual ore dilution factor.

In order to prevent any misunderstanding, the process development investigations should thus preferably start simultaneously with the drilling and mine development program. By organizing, at the early stage of the development of a new deposit, a close cooperation between the specialists in geology, mining and mineral processing, not only the procurement of the most reliable samples for metallurgical testing is greatly facilitated, but more accurate information can also be readily obtained on the amount of recoverable valuable byproducts, such as precious metals. An excellent illustration of the benefits to be drawn from such cooperation is given in Dunn's paper [1] for the development of a small low-grade porphyry copper deposit.

Obviously, the incorporation of a mineral processing expert into the team from the start of the exploration program can help orienting the exploration towards mineralized zones of greater interest on account of ore beneficiation. Cases are known, where expensive drilling programs could have been avoided by adequate testing of preliminary samples which would have shown the inaptness of the ore to be beneficiated in spite of its high grade : limonitized smithsonite deposits with zinc contents of 20 to 25 % are typical examples of such situation.

Preliminary testwork on split diamond drill cores can also help define the various types of mineralization and provide useful indications for preparing composite samples representing them. This is particularly important in the case of copper or lead ore deposits showing wide variations of the oxide to sulphide ratio.

The aforementioned remarks lead to the conclusion that a much better knowledge of the deposit and of the various responses of its ore to mineral processing may be gained by conducting the beneficiation tests on several smaller composite samples including the extreme ore types, rather than simply on one large "average" sample. It is then the metallurgist's duty to try to find out the most suitable process conditions which would best fit all cases. The amount of data collected by such testwork will be extremely valuable, not only for the designer of the plant, but even for the future mill operator. With regard to the feasibility study, it will provide the realistic margin within which the expected results could be maintained , instead of giving only one maximized grade-recovery couple of values, which probably would be reached only occasionally in the commercial plant.

While the number of individual and composite samples to submit to metallurgical testing has to be decided through a permanent dialogue between the respective responsibles for geologic development, mining plan and metallurgical process, the weight required for each sample must be determined by the metallurgist in charge of the testwork. This weight varies of course according to :

- the type of tests to be performed. Obviously, samples taken for crushing or autogenous grinding testing should be much bigger than those required for bench scale flotation tests. For autogenous grinding tests, for instance, samples should not be less than 50 tons, while 150 kg samples would be sufficient to complete the initial bench scale investigations on a porphyry copper ore [2] ;

- the type of ore. It should be kept in mind that the amount of final concentrate must be sufficiently large for assay purposes or whenever it must be submitted to subsequent testwork like roasting or leaching;

- the size of the largest particles. Reference can be made to recommended sample weights at various particle sizes, as tabulated by Taggart [3] or as calculated by Gy's theory [4]. The formula proposed by Gy is particularly useful for checking whether a given sample can be accepted or not. Moreover, the particle size of the sample must of course be also adapted to the size of liberation of the minerals, with a view to the possible applicability of gravity concentration.

In any case, the metallurgist is strongly recommended to keep a direct control of the whole sampling operation or at least to know exactly how it has been done, from where the sample comes and whether it has not been oxidized during storage or transport. He should be particularly attentive to avoid any loss (e.g. of primary slimes) or contamination (e.g. by oil, grease or other chemical products which might influence the flotation tests or even make them useless). Special care must therefore be taken in the choice of the sample container.

4. LABORATORY TESTWORK PROGRAMMING

Mineral processing flowsheet development work presents many similarities with a detective business : it requires an acute sense of observation and a thorough knowledge of all the parameters which might influence the whole process and of how this influence might be exerted. This also means that the responsible for such job must have a wide experience in the whole field of unit operations concerned, not only from a strictly technical point of view but also with a broader industrial perception of the problems, including aspects such as marketing, capital and operating costs, safety and environmental regulations. Since mineral processing involves both metallurgical operations and physical separations and is controlled by chemical and mineralogical analyses, a good comprehension of all these techniques is of course essential.

As already pointed out in the introduction, the mode of occurrence of the valuable elements in the ore is highly variable, not only from one deposit to another but even throughout any given deposit. Therefore, there cannot be any standardized method for programming the metallurgical testing. When looking for instance at the grinding, which is one of the first steps of the whole process, while the ore grindability at any given mesh is readily measured by standard methods, variations in the amenability of the ore to a given treatment such as flotation or leaching cannot be directly related to it, due to the large number of independent variables which interfere in the process.

Planning the metallurgical testwork can thus only be done after a preliminary gathering of relevant informations.

4.1. Pertinent Information to Collect

The information required concerns :

- the ore sample : its particle size, its chemical analysis and the main characteristics of its mineralization (types of valuable and gangue minerals present, and ore texture).

Very useful observations can already be made at the time the sample is received and its container opened : the visual examination of the ore may for example indicate that elimination of clay slimes by washing and scrubbing would be a beneficial first step of the process.

One should also select at this moment a few typical pieces of the ore for initial mineralogical observations (by microscopic methods and X ray diffraction) in order to identify the major mineral types and the nature of the gangue. These rough examinations will give valuable indications on the grind required for mineral liberation and help take decisions on the sampling program with regard to the types of concentration methods which should be tested and to the necessary homogeneization of the whole sample for chemical assays.

On the basis of such preliminary examinations, a decision will also be taken concerning the elements to be analyzed. These may be extended to differentiations between total metal contents and their soluble oxide forms. Depending on the type of ore, the chemical analysis should also consider the minor constituents which are known to be either favorable (e.g. precious metals) or deleterious (e.g. As, Sb, Bi).

The gangue elements should of course not be forgotten : SiO_2 (total silica and eventually quartz), CaO and MgO (total and soluble) Al_2O_3, P_2O_5, BaO and CO_2, as well as the ignition loss. The Na_2O and K_2O contents are also good indicators of the presence of clay minerals which might cause difficulties in the ore treatment by increasing the viscosity of the pulps and the consumption rate of flotation reagents. Finally, in the case of sulphide ores, total sulphur and sulphate sulphur contents should also be determined.

By combining the results of the chemical analysis with those of the mineralogical examination, a quantitative mineralogical composition of the ore can be worked out.

If any major problem arises, additional investigations should be made at this point. On the chemical side, assay methods giving an element under its various forms (e.g. lead as sulphate, as carbonate, as sulphide ...) might be very useful. On the mineralogical side, electron-probe quantitative microanalysis might be useful, e.g. for determining the iron content of the sphalerite or the composition of minerals of the sulpho-salts group [5].

Since the grind size is an essential parameter in physical methods of concentration, it is clear that it must be taken into consideration from the beginning of the ore sample characterization study. This means that after completion of any comminution stage estimated necessary for concentration testing, a detailed size analysis must be performed, including, if necessary, a sand-slime separation of the finest minus 400 mesh screen fraction. Chemical assays of the main elements of each size fraction would then permit the metallurgist to calculate the metals distribution between the various size fractions and to draw important conclusions regarding the effect of grinding.

Finally, depending on the ore type and on the observations already made, further characterization of the ore sample might still be made by densimetric analysis using heavy liquids or by magnetometric analysis at different intensities. The results of such analyses can be plotted as washability curves which describe the ideal response of the ore to the corresponding separations [6].

- specific data relative to the ore deposit location, such as water availability, climate, reagents supply, availability of quali-fied labor, transportation facilities, environmental regulations, market conditions for the end products. All these local features must be known and be taken into account in the metallurgical test-work.

Water is a particularly important parameter : its availability directly influences the flowsheet development and its quality can profoundly affect the flotation results. It is well known for ins-tance that in oxide ores flotation with fatty acid collectors, hard water would cause detrimental precipitation reactions with the calcium and magnesium ions. But drastic depression of sulphide mine-rals might also occur in the presence of some hard waters under normal flotation conditions [7].

Therefore, knowledge of the detailed composition of the availa-ble water at the future mill site is imperative. Since seasonal changes in water composition are possible, this eventuality should also be checked. Of course, for most of the metallurgical tests, the easiest procedure would be to use demineralized water, as long as the effect of the water quality change has been established. However, if this effect appears to be important, preference should be given to the systematic use of a synthetic water of composition close to that which would be expected in the commercial plant.

Data concerning the climate at the mill site are mostly useful for the design of the process plant, but may also be taken into account for the testwork, considering the effect of temperature on the flotation of some ores, particularly when less soluble long-chain fatty acids are to be used as collectors.

Choice of reagents might be restricted at some places, either for environmental reasons or because of the obligation to use local-ly manufactured products. The metallurgist should of course be aware of the existence of such constraints before starting the test-work.

Data on the availability of qualified labor should also be known, since they might lead to the choice of some processes preferably to others.

The same can be said for the transportation facilities for evacuating the final products to the market. If there is a local market which can absorb the production of the plant, a process giving a lower grade product with a higher recovery may be preferred and by-products of lower value might also bring some benefit and contribute to the viability of the project. Of course, in any case, the metallurgist in charge of the testwork must be aware, from the beginning, of the market requirements for the various categories of products he plans to produce. The better he knows the real problems encountered by the potential customers of these products, the more chance he has to work out a process that would give the greatest benefit.

Finally, before establishing the test program, it is advisable to verify whether environmental legislative requirements might not impose restrictions on the process operations due to contaminant effluents. There are cases indeed where some very specific reagents are not allowed to be used, for instance because of the necessity to protect a fishing area located in the vicinity of the plant site.

4.2. Planning the Metallurgical Testwork

When all the aforementioned data have been collected and analyzed, a rapid survey should be made of the technical documents pertinent to problems of similar nature and including not only published materials but also antecedent reports available in the laboratory own files.

With all this information in mind, the metallurgist can then define his position for proposing a program of tests to be performed. At this point, he must even be able to predetermine already the type of flowsheet which appears to be the most appropriate or at least to select a limited number of processes that should be taken into consideration during the testwork.

For example, he can already determine whether there might be any interest to include a preconcentration operation in the flowsheet, e.g. by washing out the primary slimes or by heavy media separation or any other gravity concentration method [8]. Taking such decision at this point is important, because if preconcentration operations are to be considered in the test program, they introduce more complications into the sampling program in comparison with a flowsheet based on direct flotation. In the latter case, the whole sample might indeed advantageously be crushed directly to minus 10 mesh and, after homogeneization, be used as reference feed for the various wet grinding and flotation tests.

For such crushing, one should by preference use a jaw crusher and a roll crusher in close circuit with a 10 mesh screen in such a way as to avoid overgrinding of the ore.

5. EXECUTION OF THE LABORATORY BATCH TESTWORK

Detailed description of testing procedures can be found in most of mineral processing handbooks. Some of these are given in the list of references [9] [10] [11] [12].

Without going into such details, it may be clearly stated that, being first well aware of the objectives of the test program, no serious work can be done without :

- a thorough knowledge of all the parameters which might have an effect on the actual operation test results. This means that each of these parameters must be adequately controlled and evaluated. For example, in a batch test for wet grinding in a given laboratory ball mill, following data should be recorded : total weight and size distribution of the ore sample, pulp dilution, total weight and size distribution of the ball charge, material used for the balls, water composition, amount and type of reagents added, mill speed, grinding time and size distribution of the sample after grinding;

- a keen observation of all details during actual operation testing. This is particularly important in flotation work, where the effect of addition of a given reagent can be readily observed on the froth aspect, like color, texture and loaded character. Many other details, such as pulp viscosity, presence of flocculated solids, natural floatability of particles before adding any collector, can also be easily visualized and serve as guiding-marks for the metallurgist. All such observations should be mentioned in the laboratory record book. They will help to establish experimental conditions for subsequent tests;

- a correct evaluation of test results. This first of all implies that sufficient results must be known for each test, both qualitatively and quantitatively. For instance, for a good evaluation of a laboratory flotation test, the froth should be removed in several increments, each separate concentrate fraction and the tailing being then dried, weighed and assayed, so that material and metal balances may be established and cumulative results be plotted on a graph. If the recalculated feed grade differs too much from the assay value of the main ore sample, it is preferable to discard such test results from the evaluation.

Many methods have been proposed for evaluating mineral processing batch tests results [13] [14] [15].

Some were even specially developed for specific operations or materials tested : coal washing for example led to the development of the well known Tromp's theory [16] and of the Mayer's method of presenting the separation results [17].

In ore flotation,the most convenient method for comparing tests results consists in the comparison of the cumulative metal recovery curves plotted vs. the cumulative concentrate grade curves. Another type of graph can be used by plotting the cumulative weight percent of solids in the froth as abscissae and respectively the cumulative metal recovery and the cumulative concentrate grade as ordinates. Such method of course implies the use of multiple-increment tests and, consequently, also requires a larger number of assays; but it gives very useful informations on the detailed development of the entire flotation process. By applying this method, there will be no difficulty to find out, with a minimum number of tests, the most adequate experimental conditions. For instance, at the rougher flotation stage, il will readily permit to determine the effects of grinding, conditioning time, pH, types and amounts of reagents and flotation time.

But a correct evaluation of tests results also means that interpretation must be given to eventual difficulties one might observe in getting sufficiently good results, either in overall metal recovery or in the grade of concentrate.

In such cases, detailed mineralogical examinations of separate concentrate fractions and (or) of the flotation tails will most often provide the explanation. In fact, this is also the most favorable occasion for getting the exactly and actually desired information on the mineralogical composition of the ore. Quantitative textural analysis and electron microprobe analysis are particularly useful at this point [5] [18] [19] [20]. A complete size analysis of the flotation tails or even a simple cycloning, combined with chemical assays of the size fractions, may also provide effective guidance for further testing.

The whole batch testwork must be conducted up to the point that a complete flowsheet can be derived. It therefore includes not only rougher separations, but also all the necessary cleaning stages in order to produce the desired final products.

With complex ores, such as polymetallic sulphide ores, mixed sulphide and oxide ores or ores containing several valuable non-metallic minerals, a wide variety of flowsheets may be considered, based on selective, semi-bulk or bulk separations. They may include regrind operations or combinations of several physical concentration methods and metallurgical operations such as leaching or roasting.

Although the initial test program may have been limited to the
most currently used methods for treating the type of ore concerned,
the development of the final flowsheet must actually be dictated,
step by step, by a correct interpretation of all the observations
accumulated at each stage of the testwork. Of course, the most
simple, straight forward, process should always be tried first.
Therefore, any possibility of removing the values as soon as they
are liberated should be looked at carefully and the most current
flotation reagents should always be tested before any other more
expensive one.

The final flowsheet(s) to be retained must present the detailed
material and metallurgical balances for each operation with reference to the raw ore. The metallurgical balances must of course be
established for the various metals concerned and for any element
which might have a significant effect on the market value of the
final products.

6. CYCLE TESTS

Since the final flowsheet derived from the batch testwork leads
to the production of several intermediate products, such as cleaner
tailings, which, in a commercial plant, would be recirculated, it
is essential to determine the upgradeability of these products when
reintroduced at the appropriate point of the flowsheet. It might
indeed happen that some of the recirculated products would build
up and not be recovered : they should then be submitted to a separate treatment.

Therefore, a first and easy approach might consist in separate
testing of the products to be recirculated under the same conditions
as those previously established at the point of reentry into the
main circuit.

However, a complete cycle test would be preferable. This starts
with a first batch test, using the conditions previously determined
during the main batch testwork. Several cleaner concentrates and
cleaner tails are produced, as well as a final rougher tail.
A second batch test is then performed on a new sample of the ore,
of the same weight, but as the test goes on, cleaner tailings from
the first batch tests are added at the appropriate points. This
operation is repeated several times. If a total of (n) tests has
been performed and if (a) is the number of cleaning stages in each
test, following products will be available at the end of the last
test : (n) cleaner concentrates, (n) final rougher tailings and
(a) cleaner tailings from the last test. All these products are
weighed and assayed.

The whole flowsheet with its recirculation cycles is then considered as a black box with one inlet (the raw ore feed) and several outlets (the cleaner concentrates and the final rougher tailings). The metal distributions are then calculated for each of the n tests on the basis of the outlets and a fictive feed grade is calculated by dividing the sum of the weights of metal in the outlets by the weight of the inlet.

When the equilibrium has been reached, the amount of metal introduced into the circuit is entirely divided between the various outlets; this means that the calculated fictive feed grade must then correspond to the actual feed grade.

The fictive feed grade values obtained at each stage of the cycle test are then plotted as ordinates and the metal recoveries of the final concentrates as abscissae. By joining the points in the order of the successive cycles, a series of straight segments is obtained : any point of intersection of these segments with the horizontal whose ordinate corresponds to the actual feed grade are equilibrium points. This finally permits to determine all the recoveries, weight percent and grades of the concentrates and tailings at equilibrium.

An example of data obtained from a flotation cycle test consisting of two successive flotation operations involving respectively one and four cleaning stages, is given in Table 1 where I and II refer to the respective flotation operations.

Results of this test are illustrated in Fig. 1 : points A and B correspond to equilibrium values.

Since actual cycle flotation tests may sometimes fail to reach a steady state, methods of calculation have been developed for predicting the results of such tests from batch bench-scale test data [21] [22].

Test N°	Sample	Wt (g)	% Sn	Sn Wt (g)	Sn Recov. (%)
1	final conc. I	23.7	0.54	0.13	1.6
	final conc. II	52.7	13.31	7.01	86.7
	final tail	410.7	0.23	0.94	11.7
	new feed	750	1.08	8.08	100
2	final conc. I	34.9	0.66	0.23	2.3
	final conc. II	47.4	17.99	8.52	85.0
	final tail	531.1	0.24	1.27	12.7
	new feed	750	1.34	10.02	100
3	final conc. I	59.7	0.88	0.52	7.1
	final conc. II	30.2	17.74	5.36	73.4
	final tail	525.6	0.27	1.42	19.5
	new feed	750	0.97	7.30	100
4	final conc. I	32.2	0.61	0.20	1.0
	final conc. II	85.1	18.02	15.34	77.7
	final tail	752.1	0.56	4.21	21.3
	new feed	750	2.63	19.75	100
5	final conc. I	35.5	0.61	0.22	1.1
	final conc. II	131.3	13.33	17.50	90.6
	final tail	500.0	0.32	1.60	8.3
	new feed	750	2.58	19.32	100
6	final conc. I	39.2	0.67	0.26	1.7
	final conc. II	111.9	12.12	13.56	89.8
	final tail	538.4	0.24	1.29	8.5
	new feed	750	2.02	15.12	100
7	final conc. I	49.9	0.79	0.39	2.7
	final conc. II	112.3	10.43	11.72	82.5
	final tail	584.0	0.36	2.10	14.8
	new feed	750	1.89	14.21	100
8	final conc. I	37.0	0.66	0.24	1.5
	final conc. II	140.7	10.04	14.13	86.3
	final tail	531.6	0.35	2.00	12.2
	new feed	750	2.18	16.37	100
9	final conc. I	30.1	0.48	0.14	1.0
	final conc. II	114.2	10.21	11.66	86.0
	final tail	552.1	0.32	1.77	13.0
	new feed	750	1.81	13.57	100
10	final conc. I	41.6	0.77	0.32	2.8
	final conc. II	72.3	11.02	7.97	71.0
	final tail	625.4	0.47	2.94	26.2
	new feed	750	1.50	11.23	100
	clean. tail I	54.7	1.78	0.97	
	1st clean. tail II	156.1	0.87	1.36	
	2d clean. tail II	337.5	1.54	5.20	
	3d clean. tail II	114.1	1.85	2.11	
	4th clean. tail II	60.9	3.66	2.23	
	Total outlets	7596.1	1.93	146.85	

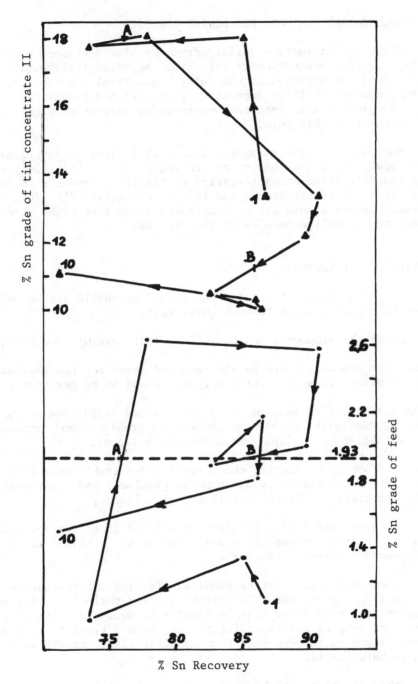

Figure 1. Example of Cycle Test Results of
Cassiterite Flotation.

7. COMPLEMENTARY TESTWORK FOR DESIGN PURPOSES

With the information derived from the aforementioned cycle tests, a better evaluation of the actual metallurgical results which should be expected in the future commercial plant can be made. It must however still be completed by several additional tests, in order to provide some important engineering data which might not yet be known at this point.

Such data are for example : Bond work indices to determine the power needed to grind the ore to the required size [23] [24], settling and filtration characteristics of the final products to determine the required thickening and filter unit areas [25] [26]. These characteristics should not be overlooked since they might strongly affect the overall economics of the process.

8. PILOT PLANT TESTWORK

Deciding whether a pilot plant should or should not be undertaken is a problem open to much controversy.

Among the arguments against pilot plant testing, one can quote :

- the high tonnage of ore sample required (even for the smallest pilot plants with capacities between 50 and 80 kg per hour);

- the difficulty of keeping the circuit under stable operating conditions (particularly due to pumping problems) and consequently of establishing reliable metallurgical balances;

- the difficulty of keeping the circuit under stable operating conditions (particularly due pumping problems) and consequently of establishing reliable metallurgical balances;

- the eventual difficulty of hiring qualified labor for operating the pilot plant around the clock, when it is installed in an independant research centre;

- the fact that a much better knowledge (easier interpretation and correction) of the possible effects of feed variations will be gained at a much lower cost by laboratory batch testwork on different ore samples corresponding to the various types of mineralizations found in the deposit, than by pilot plant testing of one large bulk sample.

Other arguments are however in favor of pilot plant testing. They refer to :

- the possibility of producing sufficiently large quantities of

material for subsequent metallurgical processing experiments
(each batch roasting test, for instance, requires a minimum weight
sample of 15 kg, but a 3 to 5 tons sample would be necessary for
a pilot scale test);

- the possibility of checking the technical feasibility of the whole
  process, particularly when there is no reference available of
  commercial use of the process for similar types of ores;

- the possibility of determining more readily and more accurately
  all the recycle effects, regarding solutions as well as solids;

- the possibility of testing new types of equipment.

9. FINAL RECOMMENDATIONS

Any owner or discoverer of a mineral deposit is hopeful that
it might sooner or later be developed into a promising viable project. Consultants are then requested to give their opinion. Normally these are geologists, and preferably economic geologists.

If the opinion of these specialists is not unfavourable, every
attempt is made to collect more information by further exploration
work, so as to be able to prepare a preliminary feasibility study.
Step by step, additional data will be gathered, which make the
project progress.

Already at these very early stages of the project, it is essential to have the geologist working in close association with a mineral processing laboratory, for the reasons shown in this paper.

Such mineral processing laboratory should have at its own
direct disposal all the facilities required for making metallurgical testwork, including chemical assays and mineralogical analyses.
It should also have a well organized service of technical documen-
tation based on most of the published books, periodicals and pre-
prints pertinent to the concerned field, and on its own internal
reports.

The organization of this laboratory must be such that mineralogists, chemists, geologists and metallurgists are used to working
together as members of a team. In each category, there should be
people with a few years experience of work in the industry.

University laboratories, where the aforementioned requirements
are filled, are in a particularly good position for acting as
referees, as long as they keep their independance, specially towards
any equipment manufacturing company.

218

As also shown in this paper, mineral processing testwork programming is, per definition, a very delicate task. It should therefore be planned, step by step, in close association with the client. Frequent meetings should be organized between all the people concerned. Such recommendation is essential and gives the best guaranty for both parts to find the most efficient solution to mineral processing flowsheet development.

REFERENCES

1. Dunn, P.G. Geologic Studies During the Development of the
   Copper Flat Porphyry Deposit. Mining Engineering, 36 (1984)
   151.

2. Smith P.R. and Butts G.E. Design of Mineral Processing Flowsheet.
   Case History, in A.L. Mular and R.B. Bhappu, (eds.), Mineral
   Processing Plant Design (New York : SME/AIME, 1980),
   pp. 860, 863.

3. Taggart A.F. Handbook of Mineral Dressing (New York : Wiley,
   1948), section 19 : 22-23.

4. Gy P. L'Echantillonnage des Minerais en Vrac, vol. 1 and 2
   (Saint Etienne : Revue de l'Industrie Minérale, 1967 and 1971).

5. Evrard L. and De Cuyper J. La Microsonde Quantitative à Balayage
   comme Méthode d'Investigation en Minéralurgie. Industrie Miné-
   rale-Minéralurgie, 1 (1978) 2.

6. Fournol H. Courbes de Lavabilité et Contrôle du Lavage. Application aux Minerais et Minéraux. Revue Industrie Minérale, 47
   (1965) 11.

7. Van Lierde A., Van Hauw J.M., Evrard L. and De Cuyper J.
   Flotation of a Zinc Iron Sulfide Ore in Water Saturated with
   Calcium Sulfate, in S. Castro and J. Alvarez, (eds.), Avances
   en Flotación, vol. 4 (Chile : Universidad de Concepción, 1979),
   pp. 45-56.

8. Morizot G. and Lafosse J. Preconcentration.
   Industrie Minérale : Les Techniques, 1 (1981) 81.

9. Taggart A.F. Handbook of Mineral Dressing Ores and Industrial
   Minerals (New York : Wiley, 1948), section 19 : 1-208.

10. Pryor E.J. Mineral Processing (Amsterdam : Elsevier, 1965),
    pp. 600-633.

11. Macdonald R.D. and Brison R.J. Applied Research in Flotation,
    in D.W. Fuerstenau, (ed.), Froth Flotation 50th Anniversary
    Volume (New York : Soc. Mining Engineers AIME, 1962), pp. 298-327.

12. Coleman R.L. Metallurgical Testing Procedures, in A.L. Mular
    and R.B. Bhappu, (eds.), Mineral Processing Plant Design (New
    York : SME/AIME, 1980) pp. 144-182.

13. Mitchell W., Sollenberger C.L. and Kirkland T.G.
    Trans. AIME, 190 (1951) 60.

14. Black K.G. and Faulkner B.P. Evaluation of Batch Flotation Results by Multiple Linear Regression. Trans. AIME, 252 (1972) 19.

15. Jowett A. Formulae for the Technical Efficiency of Mineral Separations. International Journal of Mineral Processing, 2 (1975) 287.

16. Tromp K.F. Neue Wege für die Beurteilung der Aufbereitung von Steinkohlen. Glückauf, 73 (1937) 125-131, 151-156.

17. Mayer F.W. Glückauf 86 (1950) 498-509. A new Curve Showing Middling Composition. Glückauf, 86 (1950) 498.

18. Bérubé M.A. La Minéralogie Appliquée au Traitement des Minerais. CIM Bull., 75 (1982) 121.

19. Henley K.J. Ore-dressing Mineralogy. A review of Techniques, Applications and Recent Developments. Spec. Publ. Geol. Soc. S. Afr. 7 (1983) pp. 175-200.

20. Jacquin J.P. and Gateau C. Développements de la Minéralogie Appliquée au Traitement des Minerais. Industrie Minérale, 66 (1984) 172.

21. Agar G.E. and Kipkie W.B. Predicting Locked Cycle Flotation Test Results from Batch Data. CIM Bulletin, 71 (1978) 119.

22. Cutting G.W. Process Audits on Mineral Dressing Processes : Their Generation and Practical Uses. Warren Spring Laboratory Report N° LR 271 (1978).

23. Bond F.C. Crushing and Grinding Calculation. CIM Bulletin, 46 (1954) 466.

24. Yap R.F., Sepulveda J.L., and Jauregui R. Determination of the Bond Index Using an Ordinary Laboratory Batch Ball Mill, in A.L. Mular and G.V. Jergensen, (eds.), Design and Installation of Comminution Circuits (New York : SME/AIME, 1982), pp. 176-203.

25. Miller S.A. Liquid-Solid Systems, in R.H. Perry and C.H. Chilton, (eds.), Chemical Engineers' Handbook (New York : Mc Graw-Hill, 1973), Section 19 : pp. 44-46, 60-61.

26. Leenaerts R. Essais de Laboratoire, in Société Belge de Filtration, (edr.), La Filtration dans l'Industrie des Procédés, Part V : Essais de Filtration, Sélection et Spécification du Matériel (Louvain-la-Neuve, Belgium : Soc. Belge Filtration, 1977) pp. 43-105.

MASS BALANCE EQUILIBRATION

José RAGOT, Mohamed DAROUACH, Didier MAQUIN

Laboratoire d'Automatique et de Recherche Appliquée
CRAN - ERA 905 et CRVM - L. 235
Rue du Doyen Marcel Roubault
54500 Vandoeuvre (France)

It is common practice to correct the measurements taken from a process so that random or gross errors can be accounted for and reduced. Data reliability is crucial for classical operations such as regulation, control, optimization, identification, and can be secured through a balance equilibration algorithm. Historically, balance computation proved important in determining yield accounting, in allocating costs and in recording information upon which future action can be based. From the planning point of view, balance determination is important in process design, sensors location production planning.

The general diagram of a process caracterization reveals that data validation is an essential part of the control sequence (figure 1).

1. THE ORIGIN AND CLASSIFICATION OF ERRORS

Practice has shown that measurements carried out on any process only partially verify the equations derived from its mathematical modelling. This observation can be justified as follow :

Basis :

- A model is no more than an approximation of the actual process. Its structure and parameters may have been selected or estimated from restrictive hypotheses and the computational methods may induce a certain amount of bias.

222

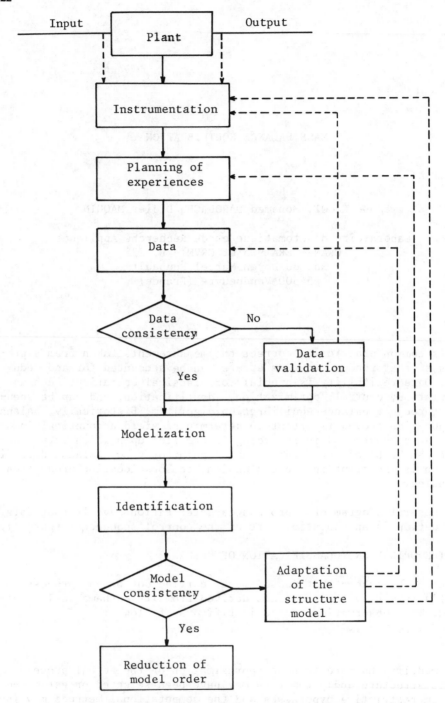

Figure 1 - Process caracterization

- Measurements carried out on the process variables do not present an accurate picture of the actual magnitudes.

In the following paragraphs only errors of the second type will be considered ; in effect we will argue that a balance equation is not derived from an approximation but that, on the contrary, it translates a physical reality.

The principle of measurement :

Measuring a magnitude G consists in establishing two reference points E_i and E_j such that :

$$E_i < G < E_j$$

In practice, this direct method is rarely used. A sequence of measurements is used instead such that this comparison is made following a more or less complex transformation of the magnitude to be measured. Thus a certain number of error factors are brought into evidence, from which we may isolate the following :

- Random factors : during the time required to effect the measurement certain magnitudes (e.g. power supply, ambient temperature, field of electrical interference, etc ...) may fluctuate in an haphazard fashion ;

- Accidental factors : either the operator or the equipment itself may sustain some temporary disorder ;

- Routine errors : ageing, maladjustment, drift or some constant external factor that not been allowed for when calibrating the measuring device may introduce a regular discrepancy whose influence may not necessarily be known.

Figure 2 shows the three types of error based on a single example : the process involved is a production unit idstinguished by its total imput flow-rate Q_1 and total output flow-rate Q_2, which are assumed to have been measured with two distinct measuring devices operating continuously.

The diagram (Q_1, Q_2) shows the relative position of the cluster of experimental points and the hypothetical curve (the first bisector of the axes).

- Random errors are pinpointed by the distance separating the experimental points from the line that "best centres" the cluster of points.

- Accidental errors are distributed as points which appear to be distinctly "more remote" from the mean than the other points.

224

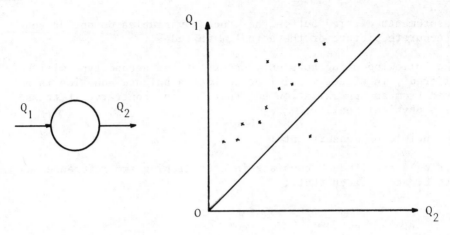

Figure 2 - The types of measurement errors

- Routine errors affect the measurements of Q_1 and/or Q_2, still following the same trend as shown in figure 2 (with this particular
  example it is impossible to predict whether the routine errors on
  measurements Q_1 and Q_2 are positive, real or negative).

2. THE NEED FOR A SYSTEMATIC APPROACH TO DATA VALIDATION

The following examples have been choosen to point out the differents
problems involved in data validation :

- type of measurement errors,

- detection of errors,

- correction of measurements,

- missing data,

- estimation of parameter,

- hierarchical validation.

Example 1 : A simple circuit shown schematically in figure 3 consists of 1 node with 3 branches such as a cyclone, a distributor, a
flotation cell ... If Q_i are solid ore flow rates and x_{ij} is the
particle size distribution, typical equations of material conservation are :

$$Q_f - Q_o - Q_u = 0 \qquad (1)$$

for the total solid mass, and

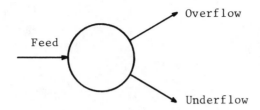

Figure 3 - **A distributor** node

$$Q_f \, u_{if} - Q_o \, x_{io} - Q_u \, X_{iu} = 0 \qquad i = 1, \ldots n \qquad (2)$$

for the solid mass per perticle size interval, where n is the number of size interval.

A typical material balance problem could be stated as follows : x_{ij} having been measured – What are the values of Q_i ? – How consistend are the measured data ?

The material conservation equation can be written :

$$x_{if} - (1 - r) \, x_{io} - r \, x_{iu} = 0 \qquad (3)$$

where r is the circulating load ratio $= \dfrac{Q_u}{Q_f}$

Obviously r can be estimated by :

$$r = \frac{x_{if} - x_{io}}{x_{iu} - x_{io}} \qquad i = 1, \ldots, n \qquad (4)$$

In the case where n   1, the measurements are said to be redundant ; i.e. there is more information than needed for circulating load ratio. The impossibility of exactly solving (3) for r is a consequence of the errors which have occured in the measurements ; the problem of mass balance is due to the fact that the data are redundant and not error-free.

Table 1 illustrates inconsistencies in data collected in a hydrocyclone classifier. An examination of the results (circulating load ratio and efficiency curve) clearly suggests that the complete set of data is suspicious.

Size (μ)	Feed (a_i %)	Overflow (o_i %)	Underflow (μ_i %)	Circulating load ratio (r)	Efficiency curve (C_i)
> 710	0,82	0	2	0,88	0
500-710	1,58	0	3,65	0,87	0
350-500	3,09	0	9,50	0,90	0
250-350	6,65	0	25,44	0,92	0
177-250	10,36	1,27	22,19	0,87	0,11
125-177	11,86	8,21	14,44	0,89	0,62
88-125	10,04	11,34	7,40	1,04	1,02
53-88	9,05	10,45	3,99	1,19	1,04
37-53	4,22	7,76	1,13	0,14	1,66
< 37	42,30	60,97	10,24	0,37	1,30

Table 1 – Test of data inconsistency

Example 2 : For the circuit given in figure 4 we will consider the solid ore flow rate Q_i and the KCl and NaCl concentrations k_i and n_i. Table 2 lists some data collected on the circuit.

As in the previous example the material balance equations can be written :

$$Q_f - Q_w - Q_c - Q_t = 0 \qquad (5.1)$$
$$Q_f k_f - Q_c k_c - Q_t k_t = 0 \qquad (5.2)$$
$$Q_f n_f - Q_c n_c - Q_t n_t = 0 \qquad (5.3)$$

The consistency of the raw data can be checked through the help of the set of equations (5) ; the residuals :

$$Q_f + Q_w - Q_c - Q_t = -652$$
$$Q_f k_f - Q_c k_c - Q_t k_t = -3113$$
$$Q_f n_f - Q_c n_c - Q_t n_t = 97$$

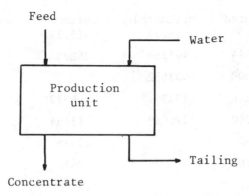

Water

Production
unit

Tailing

Concentrate

Figure 4 – A sample production unit

are clearly different from zero.

A more sophisticated test can be applied : each variable can be estimated from a direct measurement but can also be deduced from the
other measured variables through the mass balance equations. For
instance Q_f is given by :

direct measurement :   400944 t

but we can also deduce its values

from eq. 5.1 : $Q_c + Q_t - Q_w$ = 401296 t
from eq. 5.2 : $(Q_c k_c + Q_t k_t)/k_f$ = 362235 t
from eq. 5.3 : $(Q_c n_c + Q_t n_t)/nf$ = 400783 t

All such different estimates are collected in table 3.

The following comments can be adduced from an examination of the results.

	Feed (t)	Concentrate (c)	Tailing (t)	Additionnal water (w)
Flow (t)	400944	104256	1380340	1083300
KCl (%)	.2697	.8616	0.0047	–
NaCl (%)	.6068	.0382	0.1733	–

Table 2 – Measurements

All flow rate measurements are nearly equal to the estimates obtained through the total balance equation and the NaCl balance equation;
on the opposite these measurements are not consistent with the estimates deduced from the KCl balance equation. As a preliminary conclusion we should probably view with suspicious the KCl data.

	measured	estimated (5.1)	estimated (5.2)	estimated (5.3)
Q_f (t)	400944	401296	362235	400783
Q_w (t)	1083300	1083652		
Q_c (t)	104256	103904	116372	106803
Q_t (t)	1380340	1379988	3211864	1380901
k_f (%)	.2697		.2436	
k_c (%)	.8616		.9617	
k_t (%)	.0057		.0132	
n_f (%)	.6068			.6065
n_c (%)	.0382			.0391
n_t (%)	.1733			.1733

Table 3 - Test of data inconsistency

The examination of the NaCl concentrations underlines the fact that the NaCl variables are correctly measured. However it is not possible to go further in this analysis and each measurement of the KCl assay can be suspected.

3. PRINCIPLES OF MASS BALANCE COMPUTATION

The purpose of the procedure outlined here is to evaluate the flow rate of each stream of a processing circuit : flow rate of ore, concentrations, pulp solid percent, concentrations ...

The following notations will be used :

W_{ij} is the measurement of the jth variable for the ith observation.

W is the observation or measurement matrix with n rows or observations and m columns or variables formed with the elements W_{ij}.

W^* is the matrix of the true values.

$E = W - W^*$ is the error matrix.

$\hat{W}$ is an estimate of W^*

$\hat{E} = W - \hat{W}$ is the correction matrix

Usually the errors E are assumed to be realization of random variables with the following classical hypothesis :

- each random variable E_i is represented by a probability density function. The overall probability density function is written :

$$P(E/\psi)$$

- the random variables E_i are assumed to be statistically independant in regard to the different observations. We thus have :

$$P(E/\psi) = \prod_{i=1}^{n} P_i(E_i/\psi_i)$$

- the random variables E_i are assumed to be normally distributed :

$$P_i = (2\pi)^{-m/2} \left| V_i \right|^{1/2} \exp\left\{ -\frac{1}{2} \left\| E_i - \Delta_i \right\|^2_{V_i^{-1}} \right\} \quad (6)$$

where $\left| V_i \right|$ is the determinant of the covariance matrix V_i and Δ_i the
mean value of E_i.

- for the set of n observations the measurement errors for each variable are assumed to be statistically independant and of constant variance :

$$V_i = V \qquad\qquad \forall_i = 1; \ldots, n$$

where V is a m.m diagonal matrix.

- the bias error Δ_i is assumed to be constant for each variable :

$$\Delta_i = \Delta \qquad\qquad \forall_i = 1, \ldots, n$$

In terms of maximum probability the best estimator $\hat{W}$ of the true variables W^* is the one which maximizes the probability density of the
observed values W, subject to the mass balance constraints :

$$g(W) = 0$$

From the previous hypothesis, the probability density function reduces to :

$$P(W/V) = (2\pi)^{-nm/2} \prod_{i=1}^{n} \left| V \right|^{-1/2} \exp\left\{ -\frac{1}{2} \sum_{i=1}^{n} \left\| W_i - W_i^* - \Delta \right\|^2_{V^{-1}} \right\} (7)$$

where W_i^* are constrained to satisfy : $g_i(W_i^*) = 0$

Provided the covariance matrix V is known, the problem of mass balance may be stated as :

$$\min \sum_{i=1}^{n} \left\| W_i - W_i^* - \Delta \right\|^2_{V^{-1}}$$

$$\text{under} \qquad g_i(W_i^*) = 0 \tag{8}$$

Hence the Lagrange function is defined as follows :

$$\ell = \frac{1}{2} \sum_{i=1}^{n} \left\| W_i - W_i^* - \Delta \right\|^2_{V^{-1}} + \lambda_i \, g_i(W_i^*) \tag{9}$$

The stationnary conditions of first order required that :

$$\frac{\partial \ell}{W_i^*} = - V^{-1} (W_i - W_i^* - \Delta) + \lambda_i \frac{\partial g_i (W_i^*)}{W_i^*} = 0$$

$$\frac{\partial \ell}{\partial \Delta} = - \sum_{i=1}^{n} V^{-1} (W_i - W_i^* - \Delta) = 0 \qquad (10)$$

$$\frac{\partial \ell}{\partial \lambda_i} = g_i (W_i^*) = 0$$

In general the structure of the constraint g and the dimension of the problem are such that an analytical solution of the equations is not available. Consequently we have to use a gradient a Newton type iterative algorithm. In a number of cases, partitioning the criterion ℓ into sub-criteria leads to a solution that only involves a few steps of simple computations : a multiple lower level calculation is used to solve the subcriteria and a single coordination level will force the common variables of the subcriteria to take the same values.

4. APPLICATION TO A GRINDING-CLASSIFICATION PROCESS

The pilot plant of the "Laboratoire d'Automatique et de Recherche Appliquée" at the "Centre de Recherches sur la Valorisation des Minerais" of Nancy (France) is composed of :

- a ball mill (4) with its feed circuit (conveyor-belt, happer),
- a sump (8) to collect the output pulp of the ball-mill,
- a pump which control the flowrate of the input hydrocyclone,
- an hydrocyclone to classify the ore particles.

Figure 5 describes these elements and their connexions. The position of the sensors is indicated (electromagnetic flow-meters for water, weight-gauge for ore, gamma-gauge for density).

Actually the points 5 and 11 are not equiped with sensors ; the measurements at these points are obtained manually.

In the foregoing Q denotes a flow, d a density ; the subscript i refers to the "channel" i. All the defined variables obey the balance equations which are due to the conservation of the total mass and of the total volume for each sub-system : ball-mill, sump and hydrocyclone.

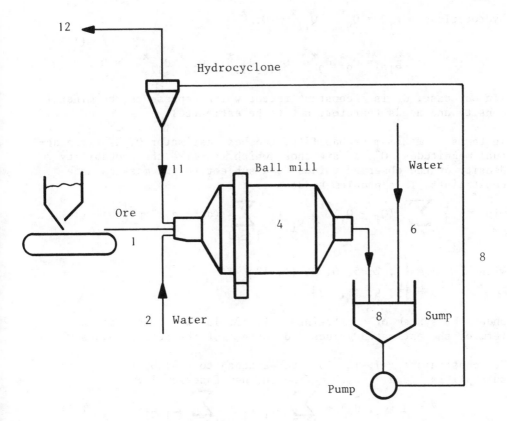

Figure 5 - Grinding-classification process

The true value Q_i^*, d_i^* (i = 1, 2, 5, 6, 9, 11, 12) satisfy the balance equations expressed in terms of volumetric flow-rate and mass flow-rate. For each sub-system we have :

ball-mill

$$f_4 = \frac{Q_1^*}{d_1} + Q_2^* + Q_{11}^* - Q_5^* = 0 \qquad (11)$$

$$g_4 = Q_1^* + Q_2^* + Q_{11}^* d_{11}^* - Q_5^* d_5^* = 0$$

sump

$$f_8 = Q_5^* + Q_6^* - Q_9^* = 0 \qquad (12)$$

$$g_8 = Q_5^* d_5^* + Q_6^* d_6^* - Q_9^* d_9^* = 0$$

hydrocyclone $\quad f_{10} = Q_9^* - Q_{11}^* - Q_{12}^* \qquad\qquad\qquad = 0$

$$\tag{13}$$

$$g_{10} = Q_9^* d_9^* - Q_{11}^* d_{11}^* - Q_{12}^* d_{12}^* = 0$$

(In our case, d_1 is a constant factor which represents the mineral density and needs therefore not to be estimated).

In terms of maximum probability, the best estimator $\hat{Q}$, $\hat{d}$ of the actual magnitudes Q^*, d^* are those which maximize the probability density of the observed values Q, d subject to the mass balance constraints. This problem reduces to :

$$\min \Phi = \frac{1}{2} \sum_{i \in I} (\hat{Q}_i - Q_i)^2 \, p_{Q_i}^{-1} + \sum_{j \in J} (\hat{d}_i - d_i)^2 \, p_{di}^{-1} \tag{14}$$

with
$$I = \{1, 2, 5, 6, 9, 11, 12\}$$
$$J = \{5, 9, 11, 12\}$$

under the system of constraints (11, 12, 13) which is written in term of the estimated values $\hat{Q}$ $\hat{d}$ instead of the actual values Q^*, d^*.

The constraints (11, 12, 13) are attached to (14) by the Lagrange multipliers   and   which define the new functional :

$$\ell = \Phi (\hat{Q}_i, \hat{d}_i) + \sum_{l \in L} \lambda_1 f_1 + \sum_{l \in L} \mu_1 g_1 \tag{15}$$
$$L = \{4, 8, 10\}$$

The conditions of first order stationarity for ℓ are :

$$P_{Q1}^{-1} (\hat{Q}_1 - Q_1) \qquad + \lambda_4/d_1 + \mu_4 \qquad\qquad\qquad\qquad = 0$$

$$P_{Q2}^{-1} (\hat{Q}_2 - Q_2) \qquad + \lambda_4 \quad + \mu_4 \qquad\qquad\qquad\qquad = 0$$

$$P_{Q11}^{-1} (\hat{Q}_{11} - Q_{11}) \quad + \lambda_4 \quad + \mu_4 \hat{d}_{11} - \lambda_{10} - \mu_{10} \hat{d}_{11} = 0 \tag{16.1}$$

$$P_{Q5}^{-1} (\hat{Q}_5 - Q_5) \qquad - \lambda_4 \quad - \mu_4 \hat{d}_5 + \lambda_8 + \mu_8 \hat{d}_5 \qquad = 0$$

$$\frac{\hat{Q}_1}{d_1} + \hat{Q}_2 + \hat{Q}_{11} - \hat{Q}_5 \qquad\qquad\qquad\qquad\qquad = 0$$

$$P_{d11}^{-1} (\hat{d}_{11} - d_{11}) + \mu_4 \hat{\hat{Q}}_{11} - \mu_{10} \hat{Q}_{11} \qquad\qquad = 0$$

$$P_{d5}^{-1} (\hat{d}_5 - d_5) - \mu_4 \hat{Q}_5 + \mu_8 \hat{Q}_5 \qquad = 0 \quad (16.2)$$

$$\hat{Q}_1 + \hat{Q}_2 + \hat{Q}_{11} \hat{d}_{11} - \hat{Q}_5 \hat{d}_5 \qquad = 0$$

$$P_{Q6}^{-1} (\hat{Q}_6 - Q_6) + \lambda_8 + \mu_8 \qquad = 0$$

$$P_{Q9}^{-1} (\hat{Q}_9 - Q_9) - \lambda_8 - \mu_8 \hat{d}_9 + \lambda_{10} + \mu_{10} \hat{d}_9 \qquad = 0 \quad (17.1)$$

$$\hat{Q}_5 + Q_6 - \hat{Q}_9 \qquad = 0$$

$$P_{d9}^{-1} (\hat{d}_9 - d_9) - \mu_8 \hat{Q}_9 + \mu_{10} \hat{Q}_9 \qquad = 0 \quad (17.2)$$

$$\hat{Q}_5 \hat{d}_5 + \hat{Q}_6 - \hat{Q}_0 \hat{d}_9 \qquad = 0$$

$$P_{Q12}^{-1} (\hat{Q}_{12} - Q_{12}) - \lambda_{10} - \mu_{10} \hat{d}_{12} \qquad = 0$$

$$\hat{Q}_9 - \hat{Q}_{11} - \hat{Q}_{12} \qquad = 0 \quad (18.1)$$

$$P_{d12}^{-1} (\hat{d}_{12} - d_{12}) - \mu_{10} \hat{Q}_{12} \qquad = 0$$

$$\hat{Q}_0 \hat{d}_9 - \hat{Q}_{11} \hat{d}_{11} - \hat{Q}_{12} \hat{d}_{12} \qquad = 0 \quad (18.2)$$

In practice an analytical solution is not possible on account of the strong linking between these 17 equations. Classical methods can be used to solve them such as a gradient or a Newton algorithm, linearisation, or direct iteration. Typical input data and results obtained from a standard program are given in table 4.

Stream	Measurements		Estimations	
	Flow	Density	Flow	Density
1	153	2.7	138.8	2.7
2	153	1	114.6	1
5	899	1.387	900.3	1.404
6	630	1	630.5	1
9	1538	1.317	1530.8	1.238
11	702	1.356	734	1.377
12	768	1.229	796	1.110

Table 4 - Measurements and estimations

5. EXTENDED BALANCE EQUILIBRATION

In classical and previous studies the problem of data validation is solved through the least squares solution constrained by linear or non linear balance equations. In practice the application of such methods is not adapted to large-scale and complex plants. This feature motivated Vaclavek [39] to attempt to reduce the size of the least squares problem by classifying the process variables into measured and unmeasured variables. More recently, Ragot [25] and Hodouin [16] pointed out the usefulness of decomposing the process in terms of macroscopic and microscopic variables. In a later work Maquin [21] emphasizes the advantage of a decentralized mass balance equibration obtained through a typological analysis of the process network.

In keeping with motivations listed above we will now survey :

- the decomposition of the process flowsheet by classifying the variables into measured, unmeasured, overdetermined, just determined.

- the decomposition of the process flowsheet into quite unconnected subprocesses by using a structural or a typological analysis.

- the decomposition of the process flowsheet through the classification of the variables into homogeneous sets.

5.1. The variables classification and the first decomposition

In the previous section we showed that the problem of data error reduction can be solved in the case of a completely observable process. Generally but unfortunately the X vector of the variables must be partitioned into :

X_m the subvector of the measured variables

$X_{\bar{m}}$ the subvector of the unmeasured variables

Some of the unmeasured variables $X_{\bar{m}}$ can be deduced in a simple way from the measured variables according to the mass balance equations. In other words $X_{\bar{m}}$ may be partitioned into :

$X_{\bar{m}d}$ the subvector of unmeasured but determinable variables

$X_{\bar{m}\bar{d}}$ the subvector of unmeasured and indeterminable variables

Moreover some of the measured variables X_m can be "recomputed" from the balance equation and with the other measured variables. Thus X_m may be partitionned as :

X_{md} the subvector of measured and overdetermined variables

$X_{\overline{md}}$ the subvector of measured and just determined variables

Consequently this classification induce the following partitioning of the process network R into :

R_u i.e. the useful network which contains only nodes with streams belonging to $X_{\overline{md}}$, $X_{m\overline{d}}$, $X_{\overline{md}}$

$R_{\overline{u}}$ i.e. the unuseful network which contains nodes with more two streams belonging to $X_{\overline{m\overline{d}}}$

An illustration of this classification is given by the flowsheet in figure 6. It is easy to see that streams 4 and 8 can be deduced from measured streams ; more completely we have the following subsets :

$$X_m = 1, 5, 9, 11$$

$$X_{\overline{m}} = 2, 3, 4, 5, 7, 8, 10$$

$$X_{md} = 1, 3, 11$$

$$X_{m\overline{d}} = 5, 6$$

$$X_{\overline{m}d} = 4, 8$$

$$X_{\overline{m\overline{d}}} = 2, 3, 7, 10$$

Let us now define the general strategy to "automatically" obtained the above classification. The vector X of the process variables is partitioned into X_m and $X_{\overline{m}}$. Accordingly the incidence matrix M is partitioned into M_m and $M_{\overline{m}}$ (obtained by permutation of the columns of M). The balance equations become :

$$M_m X_m + M_{\overline{m}} X_{\overline{m}} = 0 \tag{19}$$

If $M_{\overline{m}11}$ notes the regular part of $M_{\overline{m}}$ we have the following decomposition for M :

$$M = (M_m \ M_{\overline{m}}) = \begin{pmatrix} M_{m1} & M_{\overline{m}11} & M_{\overline{m}12} \\ \hline M_{m2} & M_{\overline{m}21} & M_{\overline{m}22} \end{pmatrix} \tag{20}$$

As $M_{\overline{m}11}$ is regular, an equivalent form for M is :

$$M = \begin{pmatrix} M_{\overline{m}11}^{-1} M_{m1} & I & M_{\overline{m}11}^{-1} M_{\overline{m}12} \\ \hline M_{m2} & M_{\overline{m}21} & M_{\overline{m}22} \end{pmatrix} \tag{21}$$

and finaly, by partial elimination we get :

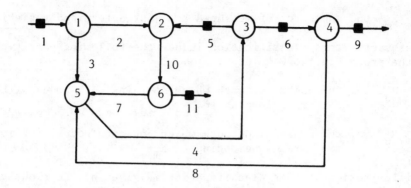

Nodes

	1	2	3	4	5	6	7	8	9	10	11←Streams
1	1	-1	-1	0	0	0	0	0	0	0	0
2	0	1	0	0	1	0	0	0	0	-1	0
3	0	0	0	1	-1	-1	0	0	0	0	0
4	0	0	0	0	0	1	0	-1	-1	0	0
5	0	0	1	-1	0	0	1	1	0	0	0
6	0	0	0	0	0	0	-1	0	0	1	-1

Figure 6 – Flow diagram for a simple system with measured streams
(■) and its associated incidence matrix

$$M = \begin{pmatrix} M_{m11}^{-1} & M_{m1} & & \vdots & I & \vdots & M_{m11}^{-1} & M_{m12} \\ M_{m2} - M_{m21} & M_{m11}^{-1} & M_{m1} & \vdots & 0 & \vdots & 0 \end{pmatrix}$$

An examination of this canonical form for the incidence matrix M
gives the classification of the variables.

In order to clarify this classification problem, let us consider our
previous example (figure 6).

Step 1 : the vector X is classified as :

X = (1   5   6   9   11 ⋮ 2   3   4   7   8   10)

Step 2 : the matrix M is written accordingly:

$$
M = \begin{array}{c}
\begin{array}{ccccccccccc} 1 & 5 & 6 & 9 & 11 & 2 & 3 & 4 & 7 & 8 & 10 \end{array}\!\!\!\leftarrow\text{Streams}\\
\left[\begin{array}{ccccc|cccccc}
1 & 0 & 0 & 0 & 0 & -1 & -1 & 0 & 0 & 0 & 0 \\
0 & 1 & 0 & 0 & 0 & 1 & 0 & 0 & 0 & 0 & -1 \\
0 & -1 & -1 & 0 & 0 & 0 & 0 & 1 & 0 & 0 & 0 \\
0 & 0 & 1 & -1 & 0 & 0 & 0 & 0 & 0 & -1 & 0 \\
0 & 0 & 0 & 0 & 0 & 0 & 1 & -1 & 1 & 1 & 0 \\
0 & 0 & 0 & 0 & -1 & 0 & 0 & 0 & -1 & 0 & 1
\end{array}\right]
\end{array}
$$

$$\underbrace{}_{M_m}\qquad\qquad\underbrace{}_{M_{\overline{m}}}$$

<u>Step 3</u> :   A permutation of the columns of $M_{\overline{m}}$ is used in order to point out its regular part :

$$
M = \begin{array}{c}
\begin{array}{ccccccccccc} 1 & 5 & 6 & 9 & 11 & 3 & 4 & 7 & 8 & 10 & 2 \end{array}\!\!\!\leftarrow\text{Streams}\\
\left[\begin{array}{ccccc|ccccc|c}
1 & 0 & 0 & 0 & 0 & -1 & 0 & 0 & 0 & 0 & -1 \\
0 & 1 & 0 & 0 & 0 & 0 & 0 & 0 & 0 & -1 & 1 \\
0 & -1 & -1 & 0 & 0 & 0 & 1 & 0 & 0 & 0 & 0 \\
0 & 0 & 1 & -1 & 0 & 0 & 0 & 0 & -1 & 0 & 0 \\
0 & 0 & 0 & 0 & 0 & 1 & -1 & 1 & 1 & 0 & 0 \\
\hline
0 & 0 & 0 & 0 & -1 & 0 & 0 & -1 & 0 & 1 & 0
\end{array}\right]
\end{array}
$$

<u>Step 4</u> :   Premultiplying the upper blocs of M by $M_{\overline{m}11}^{-1}$ gives :

$$
M = \begin{array}{c}
\begin{array}{ccccccccccc} 1 & 5 & 6 & 9 & 11 & 3 & 4 & 7 & 8 & 10 & 2 \end{array}\!\!\!\leftarrow\text{Streams}\\
\left[\begin{array}{ccccc|ccccc|c}
-1 & 0 & 0 & 0 & 0 & 1 & 0 & 0 & 0 & 0 & 1 \\
0 & -1 & -1 & 0 & 0 & 0 & 1 & 0 & 0 & 0 & 0 \\
1 & -1 & 0 & -1 & 0 & 0 & 0 & 1 & 0 & 0 & -1 \\
0 & 0 & -1 & 1 & 0 & 0 & 0 & 0 & 1 & 0 & 0 \\
0 & -1 & 0 & 0 & 0 & 0 & 0 & 0 & 0 & 1 & -1 \\
\hline
0 & 0 & 0 & 0 & -1 & 0 & 0 & -1 & 0 & 1 & 0
\end{array}\right]
\end{array}
$$

At this step in the computation, the nodes of the flowsheet associated to M are linear combinations of the nodes of the original flowsheet (due to the premultiplying matrix operation).

Step 5 : Elimination of unmeasured variables in the lower blocs of M makes explicit the canonical form of the incidence matrix :

$$
\begin{array}{ccccccccccc}
1 & 5 & 6 & 9 & 11 & 3 & 4 & 7 & 8 & 10 & 2 \leftarrow \text{Streams}
\end{array}
$$

$$
\left[
\begin{array}{ccccc|ccccc|c}
-1 & 0 & 0 & 0 & 0 & 1 & 0 & 0 & 0 & 0 & 1 \\
0 & -1 & -1 & 0 & 0 & 0 & 1 & 0 & 0 & 0 & 0 \\
1 & -1 & 0 & -1 & 0 & 0 & 0 & 1 & 0 & 0 & -1 \\
0 & 0 & -1 & 1 & 0 & 0 & 0 & 0 & 1 & 0 & 0 \\
0 & -1 & 0 & 0 & 0 & 0 & 0 & 0 & 0 & 1 & -1 \\
\hline
1 & 0 & 0 & -1 & -1 & 0 & 0 & 0 & 0 & 0 & 0
\end{array}
\right]
$$

Examining the elements of M yields the following subsets of variables :

$$
\begin{aligned}
X_{md} &= \{1, 9, 11\} \\
X_{m\bar{d}} &= \{5, 6\} \\
X_{\bar{m}d} &= \{4, 8\} \\
X_{\bar{m}\bar{d}} &= \{2, 3, 7, 10\}
\end{aligned}
$$

5.2. Partioning the network

For industrial chemical or metalurgical processes the application of mass balance does not work satisfactorily on large-scale networks. The plan therefore should be partitioned by setting up some criterion to be optimized over the set of all partitions into subsets.

This problem is a classical one in the field of system analysis and is connected with classification theory : one is given a collection of objects (the nodes of the network) on which a number of variables (the flows) have been observed and one want to express the relationships between objects in terms of a number of classes or subsets into which the objects fall.

This relationship can be adjusted by a suitable choice of a similarity or proximity measure for two objects ; an extended list is given by Chandon and Pinson. In our case the Minkowski metric appears to be sucessful in describing the distance between two nodes i and j of the network :

$$
d_{ij} = \left[\sum_{h=1}^{\sigma} W_h \left| m_{ih} - m_{jh} \right| \right]^{1/\lambda}
\tag{23}
$$

where

m_{ij} are the node coordinates

w_k are weighting factors

λ is the metric parameter (= 2 if euclidean distance is used)

As is well-known in clustering theory, the nodes or objects Q_a and Q_b are members of a cluster at level t if there exists a sequence of objects linking O_a and O_b such that the dissimilarity between each successive pair in the sequence is no greater than t. As t varies, the clusters at level t form a hierarchy ; this result implies that the clusters can be found by constructing the hierarchy either divisively or agglomeratively.

In our application we only look at agglomerative algorithm. As a first step we select the objects O_a, O_b for min (d_{ab}) ; in the second step, O_a, O_b are replaced by a cluster O_{ab} and its dissimilarity with all other objects is recomputed ; **after reinitialization these two steps** are repeated until all the objects are linked into one class.

It should be clear that we consider at each stage a current dissimilarity matrix between classes, and we amalgamate the pair of classes with smallest dissimilarity ; a rule is needed for expressing the dissimilarity between classes in term of the dissimilarity between objects in the classes.

All rules have the following property : if O_{ab} stands for the union of O_a and O_b, d (O_{ab}, O_K) is a function only of d (O_a, O_k), d (O_b, O_k) and d (O_a, O_b).

Numerous rules have been suggested ; the most commonly used are :

- single linkage $d_{fc} = \min (d_{ac}, d_{bc})$

- group average $d_{fc} = \dfrac{W_a d_{ac} + W_b d_{bd}}{W_a + W_b}$ (24)

- minimum variance $d_{fc} = (W_a + W_c) d_{ac} + (W_b + W_c) d_{bd} - W_c d_{ab}$ /

$$(W_a + W_b + W_c)$$

A comparative study of those rules is given in Gordon (13).

For the plant given in figure 7 the dendogram shown (fig. 8) is obtained sectionning the tree diagram where indicated by the dashed line yield a partition into four groups.

240

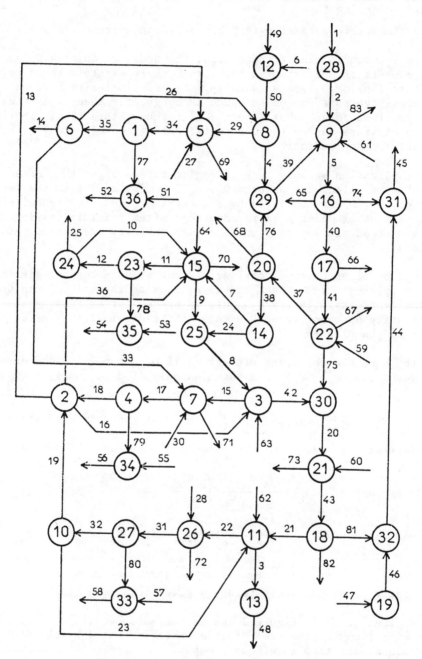

Fig. 7 - Sulfuric acid production unit circuit

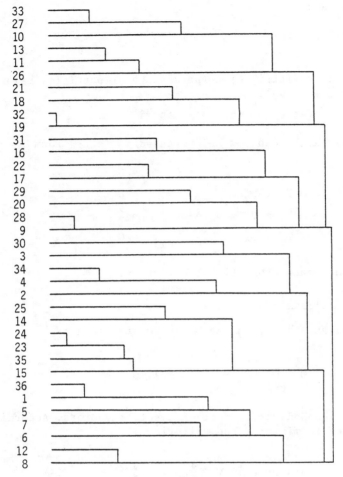

Fig. 8 - Dendogram of the circuit (Fig. 7)

Partionning the equations

Let us now consider the problem of measurement rectification for large data base systems.

In the following the flowsheet of the process will be described by means of a n x b incidence matrix M whose elements take the values 0, 1, -1 depending upon the existence and the direction of a stream at a node, with b as the number of flow streams and n the number of nodes.

The set of mass balance equations at each node are based on the total volumetric flows and on the partial flows which take the density, the granulometric distribution and the chemical species into account.

With X expanded as :

U_1 : **volu**mic flow-rate

U_2 : density

U_{3i} : granulometric distribution (percentage of particles belonging to the ith interval)

U_{4ij} : percentage of the jth specie in the ith

The balance constraints can now be expressed in matrix form :

$$
\begin{aligned}
M\,U_1 &= 0 \\
M\,U_1 * U_2 &= 0 \\
M\,U_1 * U_2 * U_{3i} &= 0 \\
M\,U_1 * U_2 * U_{3i} + U_{4ij} &= 0
\end{aligned}
\tag{25}
$$

where $*$ is the Hadamard product of two vectors.

Mass-balance equilibration

The preceding balance equation can be aggregated into :

$$
M\left(\prod_{k=1}^{L} U_k\right) = 0 \; ; \; L = 1, N
$$

where $\prod$ denotes the Hadamar product of a set of vectors. According to (8) we are led to minimize the criterion :

$$
\min_{\hat{U}_k} \Phi = \frac{1}{2} \sum_{k=1}^{N} \left\| \hat{U}_k - U_k \right\|_{V_k}^2
$$

under the constraint : (26)

$$
M\left(\prod_{k=1}^{L} \hat{U}_k\right) = 0 \; ; \; k = 1, N
$$

or equivalently :

$$
\min_{\hat{U}_k} \max_{\lambda} \ell = \frac{1}{2} \sum_{k=1}^{N} \left\| \hat{U}_k - U_k \right\|_{V_k}^2 + \lambda_k^T \, M\left(\prod_{L=1}^{L} \hat{U}_L\right) \tag{27}
$$

A necessary condition for the function ℓ to be minimum is that all the first order derivative be zero.

$$\frac{\partial \ell}{\partial \hat{U}_j} = V_j\ (\hat{U}_j - U_j) + \sum_{k=j}^{N} M^T\ (\prod_{\substack{L=1 \\ L=j}}^{k} \hat{U}_L^T)\ \lambda_k = 0 \qquad (28)$$

$$\frac{\partial \ell}{\partial \lambda_j} = M\ \prod_{L=1}^{j} \hat{U}_L \qquad\qquad = 0 \qquad (29)$$

The above system is a large scale one which is furthermore non-linear ; classical procedures can solved it but they become tedious when the number of variables gets large. In order to reduce the complexity of this problem we suggest a method based on the hierarchical calculus, that makes the following schema :

- decomposition of the objective function into a number of smaller objective functions.

- optimization of the subfunctions as if they were independent.

- coordination of the partial optimizations.

A first decomposition use a split of the variables into four subsets, each of them corresponding to the physical variables : flow-rate (U_1), density (U_2), granulometric distribution (U_{3i}), assay (U_{4ij}).

Thus, for the subscript j, in equations (28) and (29), considering all the variables $\hat{U}_i$ known except those corresponding to $i = j$, yields the solution[1]:

$$\hat{U}_j = U_j - V_j^{-1} \sum_{k=j+1}^{N} M^T\ (\prod_{\substack{L=1 \\ L=j}}^{k} \hat{U}_L^T)\ \lambda_k$$

and $\qquad\qquad\qquad\qquad\qquad\qquad\qquad\qquad\qquad\qquad (30)$

$$R_j = M_* \prod_{L=1}^{j-1} \hat{U}_L$$

$$\lambda_j = (R_j\ R_j^T)^{-1}\ R_j\ \hat{U}_j$$

and $\qquad\qquad\qquad\qquad\qquad\qquad\qquad\qquad\qquad\qquad (31)$

$$\hat{U}_j = (I - R_j^T\ (R_j\ R_j^T)^{-1}\ R_j)\ \hat{U}_j$$

The whole solution can be rewritten as :

$$\hat{U}_j = F_j\ (\hat{U}_i) \qquad \begin{array}{l} j = 1, \ldots N \\ i = 1, \ldots N \end{array} \ ; \ i = j \qquad (32)$$

or equivalently :

$$\hat{U} = \mathbf{F} \ (\hat{U}) \tag{33}$$

Although the objective function may be separated into four non-interacting functions, one for each subsystem, interaction will occur because of the interaction variables U which affect the subsystems. This coordination can be simply achieved with the help of the relaxation principle which essentially makes use of a Jacobi or of a Gauss-Seidel formulation.

Parallel resolution (Jacobi)

The whole recurrence $\hat{U}^{m+1} = F \ (\hat{U}^m)$ (34)

is used to solve all the subproblems for the iteration of rank m+1.

Sequential resolution (Gauss-Seidel)

Let us define the two associated series :

$$s = \quad 0,1, \ \dots \ m \ \dots$$
$$S = \quad s^0, s^1, \ \dots \ s^m, \ \dots \tag{35}$$

where the elements of S are choosen in the set :

$$I = \quad 1,2, \ \dots \ i, \ \dots \ n$$

The recurrence :

$$\hat{U}_j^{m+1} = F_j \ (\hat{U}^m) \qquad \text{If } j = s^m$$
$$\hat{U}_j^{m+1} = \hat{U}_j^m \qquad \text{If } j \neq s^m \tag{36}$$

with like a "chaotique" resolution of all the subproblems.

The case : $\quad s^m = 1 + m \ (\text{modulo } N)$ (37)

is known as the Gauss-Seidel algorithm.

A computer aided design system dedicated to the data validation problem

The general problem of material requirement is known in different fields under names such as material balances, mass balancing, input-output analysis, flowheet analysis, observability of process. A modular approach is particulary attractive for complex and large scale process where it is useful to evaluate numerous alternatives ; such approach identifies types of calculations which can be written as

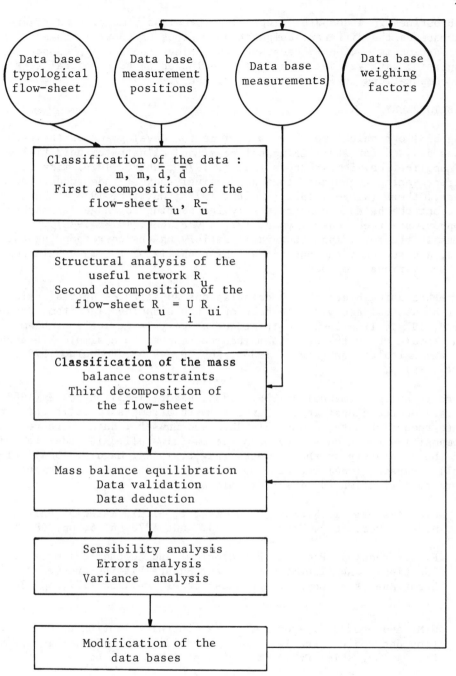

Figure 9 - A CAD system

subroutines in a computer program. As previously mentionned figure 1 summarizes the different computational steps of control process. Data validation is obviously necessary to achieve the final purpose of control ; data validation can also be looked on as a logical sequence of computations as illustrated in figure 9.

BIBLIOGRAPHY

The bibliographical references outline the development of the techniques devised for stabilizing balances, and illustrate their fields of application. Some authors (Mular, 1976 ; White, 1979) make use of direct search techniques that successively adjust independent variables. Others (Wiegel, 1972 ; Smith, 1973 ; Cutting, 1976) take into account the balance constraints by introducing Lagrange parameters. Approximate techniques (Lÿman, 1977 ; Vaclavek, 1976, 1979) yield firstly the total mass flow-rates satisfying the corresponding balances, and secondly the concentrations stabilizing the mass balances of the various components.

A method using hierarchical calculations substitutes partial balances for a mass balance with the help of a coordination algorithm (Ragot, 1978, 1979). It allows the treatment of very large-scale problems. In certain cases the algorithms can be reduced to a simple hierarchy in the calculations (Ragot, 1980 ; Aubrun, 1980 ; Hodouin, 1979 ; Sood, 1979).

More recently, Romagnoli (1979, 1983), Ragot (1983), Mercklé (1983) worked on the classification problem with attempts to split the data into redundant data, deductible data and unuseful data. Moreover decomposition techniques have been successful applied in order to reduce the complexity of the networks (Ragot, 1983 ; Maquin, 1984). Finally Brepson (1984) reported on an extension of the measurement errors rectification to linear dynamic systems.

1. J.C. Agarwal, I. Klumpar, F.D. Zybert, Simple Material Balance Model, Chem., Eng. Prog., vol. 74, June 1978, n° 6, pp. 68-71.

2. F. Albarède, A. Provost, Petrological and Geochemical Mass-Balance Equations : An Algorithm for Least Square Fitting and General Error Analysis, Comput. and Geosciences, vol. 3, 1977, pp. 309-326.

3. G.M. Anosova, S.E. Aronina, I. Ya. Shtral, Algorithmes de Correction des Débits dans les Equations de Bilan-Matière, Khim. Prom-St., Aviom. Khim. Proizvod, 1980, n° 1, pp. 22-26.

4. L. Badea, I. Belcea, Optimazarea Multicriteriala A Bilanturilor De Mareriale Ale Sistemelor Complexe, Rev. Chim. (Bucarest), vol. 31, 1980, n° 2, pp. 173-175.

5. R.F. Baldus, L.L. Edwards, Mass and Energy Balances for Complete Kraft Mills, A.I.ChE, Se. (USA), vol. 75, 1979, n° 184, pp. 56-61.

6. I. Bjerle, J.C. Berggren, H. Karlson, Chemical Process Enginee-ring, Part 1 : Material Balances, Kem. Tidskr, vol. 92, 1980, n° 5, pp. 24-28.

7. R. Bloise, C. Reinhart, J. Batina, Etablissement de Bilans Matières Statistiquement Cohérents sur des Unités Complexes, Revue de l'In-dustrie Minérale, Mars 1981, pp. 257-263.

8. J.C. Brepson, Validation de Données et Identification de Systèmes Dynamiques, Thèse de 3ème Cycle, Nancy, Juin 1984.

9. J.L. Chandon, P. Pinson, Analyse typologique, Masson Ed., 1981.

10. G.W. Cutting, Material Balances in Metallurgical Studies : Current Use at Warren Spring Laboratory, Paper 79-3, AIME Annual Meeting, New Orleans, 1979.

11. M. Darouach, Aide au Diagnostic de l'Etat des Systèmes. Validation de Mesures par Equilibrage de Bilan-Matière, Thèse de Docteur-Ingénieur, Nancy, 1983.

12. M. Darouach, J. Ragot, J.C. Brepson, Validation de Données en présence de Mesures manquantes, I.A.S.T.E.D., Lille (France), 15-17 Mars 1983.

13. M. Darouach, J. Ragot, J. Mercklé, Validation hiérarchisée de Mesures pour des Systèmes de Grande Dimension, 4th I.A.S.T.E.D. International Symposium and Course, Lugano, 21-24 Mai 1983.

14. A.D. Gordon, On the Assessment and Comparison of Classifications, Analyse de Données et Informatique, INRIA Ed., 1979.

15. D. Gun, Material Balance : Its Applications and Misapplications in Underground Coal Gasification, Report 1979, Lect. C/RI-79/05, 37 pages, Energy Res. Abstr., vol. 4, 1979, n° 22, Abstr. n.52457.

16. J. Heemskerk, Ajustement des Bilans par la Méthode des Moindres Carrés, Ind. Min., Tech., 1980, n° 2, pp. 116-118.

17. D. Hodouin, M. Everell, A hierarchical Procedure for Adjustment and Material Balancing of Mineral Process Data, Research Report 79-14, August 1979, 44 pages.

18. W. Kauschus, Compensation of Material Balances with Inaccurate Measured Values, Einsatz Vorbereitung Prozesstech., 1979, pp. 201-222.

248

19. R.G. Kheile, Solution of Material Balance Problems for Process Design, PH. D. Thesis, Purdue University, W. Lafayette, Inf., 1975.

20. D. Laguiton, Material Balance of Mineral Processing Flowsheets. Canmet. Sciences Laboratories, Division Report MRP/MSL 80-33, 1980.

21. F. Madron, Statistical Treatment of Material Balance of Separation Operations, Collec. Czech. Chem. Commun., vol. 45, 1980, N° 1, pp. 32-40, 7 ref.

22. D. Maquin, J. Ragot, J. Mercklé, Application de l'Analyse Typologique à la Décomposition des Systèmes Complexes, Journées de l'Optimisation, Montréal, Mai 1984.

23. G. Marro, R. Rossi, A. Tonieli, Appling a Two-Level Optimization Technique in Material Balance Problems, pp. 543-550.

24. J. Mercklé, J. Ragot, M. Darouach, Diagnostic assisté par Ordinateur de l'Etat d'un Réseau de Transport de Matière, 21st ISMM Intern. Symposium and Course, Lugano, 21-24 Juin.

25. A.L. Mular, R.G. Bradburn, B.B. Flintoff, C.R. Larsen, Mass Balance of a Grinding Circuit, University of British Columbia, Vancouver, B.C., Canada, CIM Bull., vol. 6, December 1976, pp. 124-129.

26. J. Ragot, Contribution à l'Extension de la Méthode des Moindres Carrés, Application à l'Equilibrage de Bilans Industriels, Thèse d'Etat, Nancy (France), Mai 1980, 308 pages.

27. J. Ragot, M. Aubrun, A. Lynch, A Useful Technique for Metallurgical Mass Balances. Applications in Flotation, 3rd Symposium IFAC on Automation in Mining, Mineral and Metal Processing, Montreal, 1980.

28. J. Ragot, J.C. Brepson, D. Sauter, Méthodologie d'Etude des Bilans-Matière. Revue de l'Industrie Minérale, 1983.

29. J. Ragot, M. Darouach, J.L. Cardini, C. Scheidt, Mise en Oeuvre d'un Programme d'Aide au Diagnostic de la Cohérence des Mesures sur des Pilotes Industriels, Intern. 83 A.M.S.E. Summer Conf., vol. 1, Nice, Septembre 1983, pp. 3-20.

30. J. Ragot, M. Aubrun, M. Darouach, Validation de Données et Equilibrage de Bilan-Matière : A Propos de Quatre Methodes, Congrès de l'IASTED, Davos, 2-5 Mars 1982, Publié dans IASTED Journal of Control and Computers, vol. 11, n° 2.

31. M. Roesch, J. Ragot, Equilibrage de Bilan-Matière et Validation de données par Fonction Spline, Congrès IASTED Measurement and Control, Tunis, 1-3 Septembre 1982.

32. J. Romagnoli, G. Stephanopoulos, On the Rectification of Measurements Errors for Complex Chemical Plants, Steady State Analysis, Chem. Eng. Sci., vol. 35, 1980, n° 5, pp. 1067-1081, 18 réferences.

33. Sbornik, Smoothing of Measured Values at Multicomponent Material Balance Calculations by the Least Square Method after Linearisation of Constraints, Sc. Papers of the Prague Institute of Chemical Technology, K 11, 1976, pp. 29-37.

34. H.W. Smith, N. Ichiyen, Computer Adjustment of Metallurgical Balances, Can. Min. and Metall., 1973, pp. 97-100.

35. M.K. Sood, G.V. Reklaitis, J.M. Woods, Solution of Material Balances for Flowsheets Modeled with Elementary Modules, The Unconstrained Case, AICHE J. (USA), vol. 25, 1979, n° 2, pp. 209-219.

36. R. Tîpman, T.C. Burnett, C.T. Edwards, Mass Balances in Mill Metallurgical Operation, 10th Annual Meeting of the Canadian Mineral Processors Canmet., Ottawa, 1978.

37. R.B. Tippin, The Application of a Computer Calculation Material Balance in Process Evaluation of a Potash Compaction, Paper 79-80, AIME Annual Meeting, New Orleans, 1979.

38. J.W. White, R.L. Winslow, Flowsheet Analysis for Mass Balance Calculation in Overdefined Metallurgical System with Recycle, Paper 79-80, AIME Annual Meeting, New Orleans, 1979, 11 pages.

39. R.L. Wiegel, Improving the Plant Metallurgical Balance, Paper 78-B-322, AIME Fall Meeting, Lake Buena Vista, Florida, September 1978, 7 pages.

40. V. Vaclavek, J. Vosolsobe, Smoothing of Measured Values at Multicomponent Material Balance Calculations by the Least Square Method after Linearization of Constraints, SB. Vys.SK., Chem., Technol. Praze, Chem. Inz., 1976, K 11, pp. 29-37.

41. I.D. Zaitsev, A. Zozulya, V.I. Shats, Principles of the Development of a Universal Algorithm for Calculation of Material and Thermal Balances of Complex Chemico-Technological Systems, Tr. N. I.I. Proekt., in-T, Asnovn. Khimii, vol. 39, 1975.

CONCENTRATING PLANT DESIGN - CAPITAL AND OPERATING COSTS

Corneille S. EK

Professor of Metallurgy, University of Liege, Belgium

ABSTRACT

Present design of concentrating plants is briefly dealt with, emphasizing the different factors which seem to be most important in these days of high labour costs and reduced availability of low-cost energy. Such items as autogenous and semi-autogenous grinding, two-stage classification, large flotation cells, pressure filtration, etc., are particularly discussed. Recent examples of small lead-zinc selective flotation plants, as well as big porphyry copper mills are described.

Methods for capital cost evaluation are surveyed according to generally accepted procedures; relationship with flowsheeting and mill design and their constant interrelations are specially pointed out.

Distribution of operating costs is indicated on the basis of processes and elements, and a few examples are fully detailed.

Reference is made to the scarce published papers which include the different above-mentioned topics.

1. CONCENTRATING PLANT DESIGN

It is not intended here to reproduce, or even summarize the large and comprehensive works recently published in excellent books [1,2], nor to simply update former contributions authored by distinguished professionals [3, 4, 5], but to briefly present thoughts on the current conceptions on plant design.

1.1. Mill Design

Mill design is one of the numerous steps that originate from successful geological explorations and are ending with an operating concentrating plant.

When the decision has come to that an orebody has to be mined and the ore processed, numerous studies have so far been carried out concerning ore reserves, mining methods, daily tonnage mined and/or processed, treatment schemes, financing, etc. Bench tests and possibly pilot-plant work have given informations for primary selection of a flowsheet which has to provide data on materials flows and processing characteristics like size distribution, pulp dilution, solid density, recycling points, etc. All this material is then passed on to the design office, where it will be committed to a project leader.

At this stage, it cannot be overemphasized that in the office entrusted with design work, being either part of the operating company, either an independent contractor, it is most successful to put together in the design team a process engineer with the project leader : this fact, well recognized by the chemical industry, is too often disregarded in the mineral industry. Furthermore, each project leader should be supplemented by an assistant project engineer permanently able to act as deputy in case of vacancy.

It must also be stressed that there are very close connections between process flowsheet, plant design and capital cost estimates : generally, one of these stages cannot be claimed to be definitively assessed until all three are settled together. This is one of the reasons why process engineer and project leader are to be partners on the job and share responsability for work completion.

Once again, it is here pointed out that mill design and capital cost estimates are advanced at same pace and that accuracy range for both features varies depending on amount of expenses and engineering time assigned to the project at a specified stage, as will be discussed in details later on.

1.2. Mill Layout

In any case, concentrating plant design is an idle task, and the statement that "no one ore has its like" can be translated into "no one mill has its like". This is particularly true as mill site is concerned, and this explains why mill layout should be considered as soon as possible in design planning.

Factors governing the selection of mill site are numerous and possibly include the following : location of mining area, schedule of mining operations, location of overburden and waste disposal areas, available municipal services (power, fuel, communication networks, rail, road, water and air transportation) and labour resources.

Geographic data comprise detailed values on general topography, hydrology and water resources, soil mechanics, meteorology (temperatures, rainfall, snowfall, mist) : careful mapping of the future site should be carried out.

When choice has been made of the ground where the mill is to be built, several options must be considered on a number of salient features :

- will a sloping terrain be used as such, or will it be levelled up or down : on one hand, advantage is taken of gravity flow for part of materials handling, but recycling will probably result in larger pumps and conveyor installations (it is however advisable to have recycle loops on the same floor); on the other hand, more power and space will be consumed for belt conveyors, but high buildings with strong foundations are avoided;

- which consideration will be given to local or regional building materials, taking into account town-planning rules and environmental regulations ?

- which part of the mill can be built in open air, or protected only by a roof, with open sides ?

- does the owner anticipate increasing concentrator capacity in the future ?

- does the owner favour a straight line or radial, more compact mill layout, in relation with previous item ?

In general, various alternatives are considered and proposed to the owner at an early design stage; when possible, a small-scale model is constructed to better visualize future mill appearance and find the correct position for the various equipment parts.

1.3. Design Features

General process data are described in Table 1. The different numerical values are bearing design factors :

- engineering factor, which is a security factor for process values, applied in the following order : weight output → volumetric flow → rounded flow;

Table 1. General Process Data	
Primary crushing	: daily feed rate feed and product size operating hours per day
Coarse ore	: total and live storage
Fine crushing	: daily feed rate feed and product size operating hours per day
Fine ore	: total and live storage
Concentrator	: daily feed rate grinding feed and product size work index classification method rougher pulp dilution and flotation time scavenger feed rate, pulp dilution and flotation time regrinding product size cleaner and recleaner circuits feed rate pulp dilution and flotation time
Thickening	: feed rate feed and underflow pulp dilution specific area $(m^2/t/d)$
Filtering	: feed rate feed pulp dilution cake moisture ($\% \ H_2O$) specific loading $(kg/m^2/d)$
Tailings disposal	: daily tonnage pulp dilution piping length and pulp flow (m/s)
Control criteria	: position and capacity of control room equipment actuation and instrumentation computer optimization

- sizing factor, which is a security factor for equipment sizing, variable according to equipment type.

In this connection, the project leader has to select the available working hours and the utilization factor for the whole mill; furthermore, he has to estimate the working rate for each equipment part (for instance, 75-78 % for primary crushing, 7 1/4-7 1/2 hours/8 hours for secondary crushing), taking possibly into account the type of material treated (i.e., for screening, 90-92 % efficiency for pebbly ores and lower figures for earthy ones).

It is well recognized now that large size equipment is more economical that small size one, advantageously reducing the number of individual pieces and consequently the building size; the statement that such a choice can result in the complete shut-down of the plant in case of failure seems to be statistically erroneous.

Large spacing between the equipment pieces increases the length of the various communication networks (conveyor belts, power and water lines, slurry pipes, etc.) and the building size, but operation supervision and maintenance are rendered much easier.

Great care must be dedicated to smoothing and shortening the movement of materials : one can here remember that a mill treating 90 000 t/d of a copper porphyry ore, is moving through the flotation plant, at a solid density of 3 and a pulp dilution of 30 %, 2,78 m^3/s, i.e. a flow of 13,7 m/s in a 20" diameter pipe.

Another possibility that cannot be neglected is the plant building according to the modular type, which takes advantage of simplicity, easier transportation in rough conditions, lower construction cost and possible ability for removing from one place and rebuilding in another one.

Experts also consider that a successful design work requires the services of an engineering firm specializing in the field, or, at least, of a well-informed consultant.

1.4. Crushing, Grinding and Classification

In recent years, the design of crushing and grinding sections of concentrating plants has recorded the most significant changes.

Today, mine and mill planning teams have to decide whether or not primary crushing should be installed in surface plant, or in pit, or underground; whether or not autogenous or semi-

autogenous grinding will be tested against more classical rod
mill - ball mill system; whether or not two - stage classification presents advantages against one-stage hydrocycloning or
classifying.

Many underground mines have decided in favour of underground
primary crusher, generally of the jaw type, due to the better
filling rate of the skips.

In-pit primary crushers set on movable platforms are gaining
more and more favour, as they permit to reduce the length of truck
haulage from the shovels and the number of trucks in operation;
ore transport from the pit to the mill is by belt conveyors.
When planning a new mine - mill complex operation, like in porphyry
copper orebodies, it is today necessary to examine and compare
many developments like in-pit crushing and conveying, skip hois-
ting, diesel trucking and trolley-assist trucking.

General use of autogenous milling is still questionned at
the present time and there appears to be no evidence of its
ability to fulfill the primary objectives set forth : ore milled
when tumbling on itself and comminuted along grain boundaries with
a high degree of liberation and a minimum energy consumption. It
seems also that small scale tests, according among others to the
Bond's Third Theory, do not give results sufficiently reliable
to dimension industrial units and that tests must be carried out
in pilot equipment.

Some project engineers claim that autogenous grinding is not
recommended for capacities lower than 10 000 t/d, or for complex
Pb-Zn sulfide ores, that its automatic control is difficult and
that a comprehensive geological and mineralogical study of the
deposit must be performed beforehand. However, one can see in
Swedish plants, for instance, autogenous mills used in many
occurences, and for different ore types, with a high degree of
automation : people in charge of those mills are generally most
satisfied.

High circulating loads are sources of many discussions; they
can show some advantages, but there appears to be trends towards
reduced circulating loads resulting from better classification,
such as provided by two-stage systems (i.e., hydrocyclone -
cone classifier or spiral classifier - hydrocyclone, or hydro-
cyclones in series).

1.5. Flotation

Flotation progress has kept pace with increasing complexity and decreasing values content of most of the ores treated today. Except for marketing of new reagents with better properties and constant improvement of the modeling of plant configuration, it must be stressed that cell design and cell size have shown tremendous changes since a few years.

Large flotation cells are now standard in many plants, whatever their production capacity, and 60 m^3 cells are used on a trial basis in Finland. At present, it is claimed that for big mine-mill operations, like some of the low-grade porphyry coppers, large flotation cells are indisputably the best choice, for many proposed advantages have been checked, confirmed and illustrated. What is still in dispute is the minimum number of cells to be set in a row in order to meet the requirements of acceptable pulp flow, residence time distribution and metallurgical flexibility. When floating complex sulfide ores in medium-size or small plants, it seems that avoiding short-circuits in a bank results in the use of smaller cells : this explains why present practice refers mainly to copper roughing-scavenging circuits, although companies like Outokumpu in Finland use big cells in lead-zinc circuits. The large mechanical cells are however challenged by some pneumatic systems like Flotaire, Davcra, Maxwell or Column flotation cells; their use is not very widespread, but satisfactory results are reported by different operators.

Furthermore, design engineers must anticipate the development, in a not too far future, of new types of flotation cells based on a better knowledge of hydrodynamics, microturbulent patterns and operation kinetics, resulting in a reduced energy consumption with equal or better metallurgy.

Cell size and cell number determination is founded on laboratory data, and the present selection of scale-up factors is sometimes based on guess more than rigorous theory, or even reliable practice. For instance, scale-up of flotation time is not an easy task and it requires a good knowledge of ore type response and cell design criteria to correctly decide upon cell number and circuit outline : in that respect, the sparkling achievements of flotation circuit modeling are noteworthy, as it will be dealt with in another chapter of this book.

Although the many aspects of the work performed by the flotation process cannot be covered at any length in this chapter, one can however emphasize some particular points :

- simplicity and flexibility are often opposite requirements, but this must be the goal set for the flotation plant by the design engineer; simplicity of the circuit permits better control of all the cell functions and easier management of the process, but flexibility is needed to cope with varying feed grades and tonnages, changing floatability characteristics, possible breakdowns, etc.; these are the reasons why roughing-scavenging sections are generally considered as a whole, without any recycling, and why many designers prefer to set 2 parallel circuits in that flotation step;

- optimization of flotation metallurgical results depends on the strict and adaptable control of all the cell functions; it is now recommended to separate the pumping and the aeration duties of the cell impeller from its truly flotative work : this explains why most of the new cell types are supercharged, allowing for a much more precise adjustement of air flow;

- in spite of the development of automatic control of the flotation process, this still requires the supervision of a trained operator; inspection and maintenance are easier when the flotation cells are installed at an adequate height above ground, with the pumps sited at a lower level;

- flotation plant lay-out has to provide for a possible expansion, as this fact has been observed so many times in so many ore treatments;

- for ores of all types, the trend is to decreasing values content and increased intergrowth; grinding is getting finer and finer, resulting in a decrease of preconcentration methods suitability (ore sorting, heavy-media separation, jigging, etc.); however, flotation itself can be used as a preconcentration method, applied in the grinding circuit (ball-mill or even rod-mill discharge) to scalp out the liberated values, as coarse as possible, before their sliming in circulating loads; unit cells have been designed to accomodate for coarser feed without too much sanding;

- big flotation cells are preferably installed step-wise, to make use of gravity influence for very large pulp flows.

1.6. Flotation Products Handling

Products of the flotation process are one or several concentrates and tailings.

Generally, the concentrates are dewatered in classical thickeners, then on filters of different types, and, possibly, in dryers when metallurgical process or transportation requirements demand a low moisture content.

At the present time, use of lamella thickeners for concentrates is uncommun (what is probably due to their very limited surge capacity), and progress in this field is mainly related to the development of new flocculants.

As far as filtering is concerned, the usual drum or disk vacuum rotary filters are challenged by new models of belt filters and, chiefly, of automatic chamber or pressure filters. In some cases, the latter type is able to cut down the moisture content to a figure low enough to discontinue drying, which is a high cost operation at today's fuel prices. The use of this type of filters has even been considered for tailings.

Tailings disposal is quite a matter : in some mining operations, tailings are likely to represent 90 to 95 % of the ore feed, i.e. up to 80 000 or 100 000 t/d. Choice of the site, construction of the dam, possible hydrocycloning, seapage troubles, water recycle, land-slide dangers, gravity flow or pumping are but a few criteria that are to be evaluated with utmost care during concentrating plant design : example of molybdenum Henderson mine and mill is at the best illustrative in that respect.

1.7. Energy Consumption

Energy usage in mineral processing was recently the subject of specific papers [6][7]. In a specific case (1 % Cu sulfide ore processed by flotation), it has been calculated that comminution and water supply can represent 60-70 % of the total energy cost, with around 50 % for grinding alone : this fact explains why so many attempts are made in order to improve the grinding – classification circuits.

However, it must be pointed out that, per ton of refined metal, energy consumption is possibly higher in beneficiation stages than in smelting and refining. Due to the present low prices of metals, leading to search for savings in all the process stages, it is mandatory to consider the whole flowsheet, from mine to refinery and adapt technology to the better economic balances.

1.8. Environmental Problems

Environmental considerations are to be dealt with at a very early stage of the plant design. In many cases, tailings disposal is the main constraint, as described hereabove, and in the future, effluent quality will probably be checked with permanent instrumentation, in order to monitor the chemical and/or biological treatment of disposed water.

However, ecological protection is possibly realized at its best when any effluent (solid, liquid or gas) can be recycled (solids as backfill in an underground mine, pond overflow as process water, etc.).

Drying of concentrates and losses during transportation can also represent pollution sources, which require consideration in due time.

1.9. Porphyry Copper Ores

Porphyry copper ores originate from huge orebodies. with very low grades; the economics of ore fruition require the mining and the processing of very large daily amounts of ore and waste.

Fine dissemination of sulfides and/or intergrowth of copper sulphides with pyrite result in the fine grinding and the flotation of the bulk of the ore : recent examples of such concentrating plants are described in specialized books and periodicals [8][9][10][11].

One can here observe trends towards increasing use of semi-autogenous or autogenous grinding (Aitik, Palabora, Bougainville, Lornex, Similkameen, Pima, Bagdad, etc.), resulting in the construction of very big shells powered by one or two electric motors. Increased diameter and lighter grinding medium could be the cause of higher energy consumptions.

Good metallurgical performance of large flotation cells, combined with lower floor surface, lower specific energy consumption and easier process control, is leading to their necessary setting up in these concentrators.

1.10. Lead - Zinc Ores

As early as 1953, a very useful and detailed paper [12] has been published by Wright on the design and construction of small lead-zinc concentrators in British Columbia; although limited in time and geographic area, this paper covered the whole field of plant design and capital cost, and contained a wealth of information. To the best of my knowledge, it has never been supplemented afterwards.

In recent years, it appears that a few changes have been introduced in the processing of lead-zinc ores :

- in many concentrators, autogenous or semi-autogenous grinding has found favour, mainly in Europe (Saint-Salvy in France, scandinavian and finnish mills);

- development of big and rich orebodies in remote areas, like Northern Canada, has led to original solutions, like the Polaris barge [13].

2. CAPITAL COST

2.1. General

It is now possible to find adequate references for capital cost estimation : some of these are very general, coping with the whole mine-mill operation [14][15][16] some are restricted to mineral beneficiation [17], and one is specific of flotation plants [18] : in the latter paper, 3 examples are worked out in details.

It seems that progress in this field has derived from a "chemical engineering" approach, based on the "unit operations" concept and from a better file-keeping and file-release in many engineering companies.

However, as already pointed out, the concentrating plants are all different, and design engineering and capital cost estimation require more time with more people than in the chemical industry. Nevertheless, one can also find in mineral process engineering the three common steps : pre-feasibility, feasibility, construction.

2.2. Prefeasibility Engineering

At this stage, which is intended to provide the means of accepting or rejecting a project, i.e. deciding in fact to put more money in an enlarged project, design and capital cost estimations are limited to a gross value that can be derived, according to the size of the project and the complexity of the flowsheet, by 2 or 3 persons working 2 to 4 weeks.

The prefeasibility project comprises :

1. project definition with indication of battery limits

2. selection of data used to prepare the plant design

3. flowsheet description with equipment listing

4. mass balances for the different principal equipment units

5. dimensioning table giving the unit flows corrected by the suitable factors

6. calculation of values permitting preliminary equipment specifications with water and electricity consumptions

7. listing of major equipment units.

At this stage, accuracy of $\pm$ 30 % is obtained by analogy with existing plants, but it is generally preferred to subdivide the plant into different sections (crushing, grinding, flotation, etc.) and to apply judgment as to corrections to be applied when process or tonnages are different.

It is very common to use the "six-tenths rule", which is written

$$\text{cost } P_2 = \text{cost } P_1 \left(\frac{\text{capacity } P_2}{\text{capacity } P_1}\right)^f$$

where f is a factor usually taken as 0,6 in chemical engineering, and maybe nearer to 0,7 in mineral processing. This factor f is a gross value, adequate for a whole mill, but it can be varied according to the selected section of the plant and to the practical experience of the project leader.

When such a project is prepared by an engineering company, it is highly possible that they draw from their knowledge of many cases to provide a quicker and safer response.

It is obvious that cost P_2 is valid in the same year as cost P_1, generally in the past. In order to estimate the cost now, cost P_2 is multiplied by one of the published cost indexes, like Engineering News Record construction index, Chemical Engineering plant construction cost index, or Marshall and Swift cost index, the latter one having several values according to the type of industry : best choice could be M-S mine/mill cost index. The accuracy of cost indexing is about 10 % over a time lapse of 5 years, and it decreases after that.

2.3. Feasibility Engineering

When a general agreement has been reached on fundamentals, a feasibility study can be ordered in view of arriving to a decision of constructing the plant, or not.

Here, the proposed process data are surveyed in details, process equipments are speficied and selected, and general l'ayout of the different sections can be drawn. At this moment, accuracy reaches $\pm$ 10 to 15 %.

The feasibility project comprises :

1. same as in 2.2

2. same as in 2.2

3. same as in 2.2, with further listing of connecting lines and instrumentation

4. same as in 2.2

5. same as in 2.2

6. calculation of values permitting complete and definitive equipment specifications, water and energy consumptions

7. determination of the water circuit lines

8. listing of the principal electric motors in order to calculate energy consumption

9. listing of equipment specifications

10. drawing of general plans, with level selection.

The work involved at this stage requires from 4 persons during 2 months up to 10-12 persons during one year.

Capital cost estimates are prepared by summing up the different items of delivered process equipment (delivered price is generally taken as 1,03 x factory cost), and multiplying this sum by a plant cost ratio that has the following values :

solid process plant   : 3,1

solid-fluid process plant : 3,6

fluid process plant   : 4,7

Another method is the plant component cost ratio method [17], where delivered equipment costs (taken from references or factory consultations) are the base of a factored summation to include major cost additions like equipment installation, piping, electrical, instrumentation, buildings, plant services, site improvements, field expenses and project management. The total cost obtained represents the fixed capital cost, and it is necessary to add a supplement (10 to 20 %) for the working capital cost, covering materials inventory, accounts receivable and available cash.

2.4. Construction Engineering

When plant construction has been settled, all the items included in the feasibility engineering project are revised, and flowsheet is fully dimensioned.

At this moment, engineering company and client have to finally freeze flowsheet and design, in order to avoid further delays and postponements.

The project now includes :

- detailed study of the process
- specification and selection of all the equipments
- detailed construction drawings
- construction planning
- start-up procedure.

2.5. Project Reports

At the end of each stage of the project, the project leader must prepare for the client a progress report which generally includes the following items :

1. Summary and conclusions

2. Position of the project in the company's development

3. Scope of the project

4. Cost estimation : fundamentals
   repartition by sections
   estimated accuracy

5. Planning

6. Economic aspects.

2.6. Examples

In U.S. $/t/d, 1980, capital costs of concentrating plants are probably comprised in the following range :

t/d	$/t/d
10 000	2 000 - 6 500
20 000	2 000 - 5 000
50 000	2 000 - 3 500
80 000	2 000 - 3 000

From 1973 values, it has been calculated that capital costs varied as follows :

10^3 t/d	M $, 1980
20	80 - 85
50	130 - 140
70	190 - 210

3. OPERATING COSTS

3.1. Introduction

Operating costs for beneficiation plants can be found in many papers dispersed in a number of periodicals, but they are not collected in a systematic way. However, published values give a rather good picture of the past and present state, and detailed methods of calculation are described in a very good manner [14][15] [19][20].

3.2. Operating Cost Distribution

Operating costs are generally distributed in direct costs and indirect costs; the latter costs include mainly all the items which are not directly related to in-plant operation, i.e. supervision, clerical, engineering, assaying, research and development, purchasing and selling, safety and security, all the supplies for the above-mentioned items, various communication expenses (telephone, postage, telex, etc.) specialized services (accounting, computing, etc.), and other administration operations.

Direct operating costs can be distributed according to processes or to elements, as indicated hereunder :

Cost distribution by processes	Cost distribution by elements
Crushing	Labor
Grinding and classification	Power
Flotation	Supplies
Thickening and filtration	Steel
Concentrate handling	Shops
Tailing disposal	Reagents
Water supply	Others
Maintenance	
Assaying	
Supervision	
Others	

In some summary forms, operating costs are divided as follows :

- Direct Milling Costs
  . operation (cost distribution by elements)
  . maintenance (cost distribution by elements)

- Indirect Milling Costs

Total amounts and amount per processed ton are indicated for the current month and for the year to date.

However, for a mill superintendent, it may be clearer to distribute the costs by processes, for operation and for maintenance, in order to get a better knowledge of the evolution of the concentration cost price.

3.3. Examples

In U.S. $/t, 1980 value, direct concentrating cost could vary as follows, according to mean general figures :

Concentrating capacity, 10^3 t/d	Concentrating cost, $/t
20	2,00
40	1,50
70	1,25

As related in the Canadian Mining Manual, one can find the following values, in U.S. $/t, 1983 :

	t/d	$/t
Copper mills	1350 - 19500	2,29 - 5,42
Copper-zinc and zinc mills	1440 - 6530	3,27 - 5,58
Copper-lead-zinc mills	900 - 9000	5,73 - 10,05

For copper porphyries, treating 20 000 to 50 000 t/d, it can be estimated that operating costs are distributed as follows :

	¢/t, 1980
Crushing	30 - 65
Grinding and classification	50 - 80
Flotation	15 - 30
Others	12 - 77
Total	150 - 270

266

1. McQuiston, F.W. and Shoemaker, R.S. Primary Crushing Plant Design (New-York, SME/AIME, 1978).

2. Mular, A.L. and Bhappu, R.B. Mineral Processing Plant Design (New-York, SME/AIME, 1978).

3. Weiss, N. and Cheavens, J.H. Present Trends in Mill Design, in N. Arbiter, ed., Milling in the Americas (New-York, AIME, 1964), pp. 11-51.

4. Salat, S.J. How planning and design know-how save money in mill construction. World Mining 26 (october 1973) 54-58.

5. Shoemaker, R.S. and Taylor, A.D. Mill Design for the Seventies. Trans. SME/AIME 252 (june 1972) 131-136.

6. Pazour, D.A. Saving energy in beneficiation and refining. World Mining 35 (april 1982) 44-48.

7. Cohen, H.E. Energy usage in mineral processing. Trans. I.M.M. (Inst. Mining and Met., Trans. C) 92 (septembre 1983) 160-164.

8. Daman, A.C. Phelps Dodge Tyrone concentrator. Deco Trefoil 35 (fall issue 1971) 9-24.

9. Otis Staples, C. Bougainville Copper Pty Limited. Deco Trefoil 38 (1974) 7-14.

10. Fahlström, P.H., Fägremo, O. and Gjerdrum, A.S. Autogenous grinding at Boliden's Aitik plant. World Mining 28 (march 1975) 42-47, (april 1975) 42-46.

11. Section 7 : Plant practice - Sulfide Minerals - Reviews of present practice at selected copper concentrators, in M.C. Fuerstenau, ed., Flotation (New-York, AIME, 1976) pp. 1027-1144.

12. Wright, H.M. Design and construction of small concentrators in British Columbia, in Recent Developments in Mineral Dressing (London, I.M.M., 1953) pp. 719-753.

13. Leggatt, C.H. Polaris - World's most northerly mine. World Mining 35 (september 1982) 46-51.

14. Straam Engineers. Capital and operating cost estimating system handbook - Mining and beneficiation of metallic and non-metallic minerals except fossil fuels in the United States and Canada (Washington, U.S. Bureau of Mines, 1978).

15. O'Hara, T.A. Quick guides to the evaluation of orebodies. CIM Bulletin 73 (february 1980) 87-99.

16. Mular, A.L. Mining and Mineral Processing Equipment Costs and Preliminary Capital Cost Estimations (Montreal, C.I.M., 1982).

17. Balfour, R.J. and Papucciyan, T.L. Capital cost estimating for mineral processing plants. Canadian Mineral Processors, 4th Annual Meeting (january 1972) 157-188.

18. Michaelson, S.D. Flotation Economics - Part II : Capital costs of Flotation Mills in D.W. Fuerstenau, ed., Froth Flotation (New-York, AIME, 1962), pp. 612-657.

19. Weiss, N. Flotation Economics - Part I : Operating costs, in D.W. Fuerstenau, ed., Froth Flotation (New-York, AIME, 1962), pp. 584-611.

20. MacDonald, R.A. Methods for Developing Estimated Costs for Concentrating and Processing Facilities, in J.R. Hoskins and W.R. Green, eds., Mineral Industry Costs (Spokane, Northwest Mining Association, 1977), pp. 163-177.

SIMULATION IN MINERAL PROCESSING

Gordon E. Agar, B.A.Sc., Sc.D.

Section Head, Mineral Processing Research, INCO, Ltd.   2060
Flavelle Blvd., Sheridan Park, Mississauga, Ont.   L5K 1Z9, Canada.

ABSTRACT

A technique for simulating a continuous flotation circuit from batch test data is described and demonstrated. Because the simulation is based on a kinetic analysis of the batch data, a method for accurately collecting rate data from a batch test is also described; and based on the kinetic results, it is shown that an optimum time can be specified for a flotation stage. Sequential application of this procedure to the successive stages of a circuit lead to an optimized flowsheet.

An empirical demonstration of these techniques on a prospect sample is included. A three stage counter current flowsheet was optimized, continuous circuit results were simulated, and finally an experimental locked cycle test was executed to demonstrate the agreement between simulated and experimental results.

INTRODUCTION

Simulation of various aspects of mineral processing operations has been practiced in one form or another for many years. Primarily, the emphasis has been on empirically based experimental methods; for instance, batch locked cycle tests, that culminate in pilot plant testing. From the observations of the difficulty encountered in commissioning new plants it might be concluded that these methods could be improved. The increased emphasis on mathematical simulation techniques and particularly the ready availability of high capacity computers has enhanced interest in this general area of mineral processing.

Grinding and classification have received the far greatest attention; perhaps because that phase of a mineral separation operation consumes much of the total energy used in a plant and correspondingly is the most expensive portion; perhaps because it is so vital that the desired liberation be achieved effectively without unnecessary size reduction of liberated particles so that the subsequent separation process can be practiced most efficiently; or perhaps it is because the topic was accessible to laboratory scale investigation without regard to chemical conditions that so affect other surface sensitive aspects of mineral processing operations. For whatever reason, the technology has been extensively developed and the population balance modelling of grinding and classification has reached the industrial application stage. Some noteworthy applications were in the evaluation of grinding aids [1] and in characterizing the performance of a centrifugal mill [2].

Other aspects of mineral processing have not received the same intensive research attention accorded to grinding. Of these, flotation is one of the most predominant and except for the work of King [3,4] and Lynch et al [5] not much is readily available in the literature.

The application of simulation procedures for separation processes such as flotation is evident when the use of batch test data is examined. Every year thousands of batch flotation tests are executed, ostensibly to evaluate samples from prospects. To be truly useful it is essential that these prospect evaluation tests yield accurate estimates of the concentrate grade and mineral recovery that would be obtained in a continuous plant operation. Obviously, some technological scheme such as simulation is required to transform the results of batch tests into the corresponding results for a continuous circuit [6]. If the basic grade-recovery data can be generated, the next requirement is a complete mass flow balance for a circuit which can be used to size the equipment to carry out the separation. Again a simulation procedure using the batch data will satisfy this need.

Fortunately, the procedure applicable to prospect samples will be equally applicable to the treatment of intermediate streams in an operating plant. With an effective simulation procedure batch testing could be used to examine process alternatives and pilot plant testing could be appropriately used to confirm the results predicted rather than to conduct investigations as is often the case now. In what follows a procedure will be described which meets these needs and some applications of the procedure will be discussed.

Once a simulation scheme is available the need for an optimization procedure becomes apparent. The accurate translation of batch test results to continuous circuit results is the first step; the next step is to generate an optimized circuit design [7]. The basic premise for the optimization is that flotation is a kinetic process which can be described by appropriate mathematical expressions. While several criteria may be used to define the optimum separation in a stage, the one preferred states that at the optimum time the flotation rates of the valuable component and the gangue are equal. For a simple two component system this is equivalent to terminating flotation when the incremental concentrate grade decreases to the feed grade. Another statement of this is to add nothing to the concentrate that is lower in grade than the feed to the separator. Simultaneous application of these two procedures, simulation of continuous results and stage optimization, provides a means for efficiently generating an optimized circuit design from batch test data.

METHODOLOGY

Simulation

A continuous circuit can be simulated from batch test results by assigning "split factors" to each of the components in a separation stage [6]. The split factor, which is illustrated in Figure 1, has been defined as the distribution of the component in the non-float portion. It was assumed that the split factors are functions of flotation chemistry, mechanical conditions and flotation time. Thus, by fixing the chemical and mechanical conditions and the time, split factors for a given separation stage will be fixed. It must be emphasized that the split factors are based on the feed to the individual separation stage, not the feed to the circuit.

Besides the other assumptions it is apparent that the assumption must also be made that the behavior of minerals in a separation stage must be independent of the solids composition of the pulp. This is a sort of Dalton's Law, applied to flotation. The independent behavior of each size fraction was demonstrated

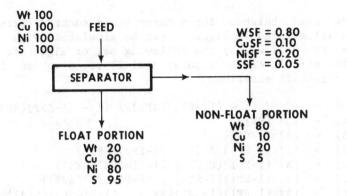

FIGURE 1 – Illustration of the definition of split factors

for ilmenite flotation where, from split factors of the various size fractions, the flotation of a variety of different size distributions could be predicted [8].

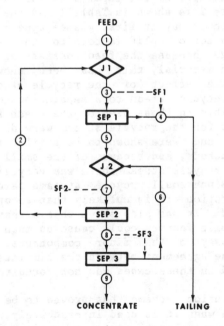

FIGURE 2 – Flowsheet of a three stage countercurrent separation circuit

The mass balance for a three stage counter current circuit that is illustrated in Figure 2 can be simulated with the split factors as defined by the following set of algebraic equations. Any other circuit can be simulated similarly with an appropriate set of algebraic equations.

$$[3] = [1] / (1 - ((1-SF1)(SF2) / (1 - (1-SF2)(SF3))))$$
$$[4] = [3] (SF1)$$
$$[5] = [3] (1-SF1)$$
$$[7] = [3] (1-SF1) / (1 - (1-SF2)(SF3))$$
$$[2] = [3] (1-SF1)(SF2) / (1-(1-SF2)(SF3))$$
$$[8] = [3] (1-SF1)(1-SF2) / (1-(1-SF2)(SF3))$$
$$[6] = [3] (1-SF1)(1-SF2)(SF3) / (1-(1-SF2)(SF3))$$
$$[9] = [3] (1-SF1)(1-SF2)(1-SF3) / (1-(1-SF2)(SF3))$$

From the mass flow of each of the components and the feed assay, the assays of each stream can be calculated. Although this procedure is straightforward and produces an exact solution for the circuit, the calculations become tedious when several size fractions are considered. For this reason a computerized version of this calculation was prepared and, to provide flexibility in examining a variety of flowsheets, an iterative procedure was used to determine the distributions rather than the set of algebraic equations. An example of the input data for the simulation of the circuit in Figure 2 is shown in Table 1 and the output is shown in Table 2. While it may at first glance appear to be a limitation to apply a given set of split factors to the entire feed to a separation stage because the feed consists of a mixture of fresh feed and recycle material, this has been shown to have minor impact [6]. The effect of the recycle stream was examined by subjecting the recycle stream to a separation under conditions as close as possible to those used on the fresh feed stream. A set of split factors for the recycle stream were designated "secondary split factors" and were shown to be similar in magnitude to the primary split factors, and because of the small mass fraction of material in the recycle stream there was very little impact on the final result. Since small recycle streams are preferred in a continuous operation it is unlikely that secondary split factors will be required. It was also shown that a change of as little as 10% in the split factors could cause as much as a 75% change in the simulated assays of the minor components. Thus, it would appear that the agreement between the simulated and experimental results observed in these cases was not fortuitous.

This simulation scheme has proven to be a valuable tool in flowsheet design when it is used interactively with the batch test program. Following batch tests a variety of circuits may be simulated. If the desired results are not achieved then adjustments may be made to the flotation chemistry or if mineralogical examination justifies it, a regrind stage may be

TABLE 1 – Computer input for a typical three stage counter current flotation separation flowsheet

```
UITKOMST U74-2
BATCH TEST 4156 LCT 12 (4149)
LOCKED
        4       1       3       9       5
CUNIFE
CU FEED ASSAY
   .170000017
NI FEED ASSAY
   .410000026
FE FEED ASSAY
  11.8800001
        1       2       1       2       3
        1       1       3       4       5
        2       2       5       6       7
        2       1       7       2       8
        3       1       8       6       9
WSF1
   .796599984
WSF2
   .608799994
WSF3
   .354300022
CUSF1
   .782999992E-01
CUSF2
   .699999928E-01
CUSF3
   .513000004E-01
NISF1
   .131299973
NISF2
   .136900008
NISF3
   .296999991
FESF1
   .615700006
FESF2
   .292999983
FESF3
   .214999974
```

TABLE 2 – Computer output of the locked cycle simulation of a three stage counter current flotation separation

UITKOMST U74-2
BATCH TEST 4156 LCT 12 (4149)

STREAM NUMBER	CALCULATED ASSAY				CALCULATED RECOVERY		
	CU	NI	FE	WEIGHT	CU	NI	FE
1	0.17	0.41	11.88	100.0	100.0	100.0	100.0
2	0.07	0.46	10.83	16.8	7.3	19.0	15.3
3	0.16	0.42	11.73	116.8	107.3	119.0	115.3
4	0.02	0.07	9.07	93.0	8.4	15.6	71.0
5	0.71	1.78	22.16	23.8	98.9	103.4	44.3
6	0.22	3.82	24.69	3.8	5.0	35.6	7.9
7	0.64	2.07	22.51	27.6	103.8	139.0	52.2
8	1.52	4.56	40.68	10.8	96.6	120.0	36.9
9	2.24	4.97	49.46	7.0	91.6	84.3	29.0

```
NO. OF ITERATIONS FOR WEIGHT =  6
NO. OF ITERATIONS FOR CU =  5
NO. OF ITERATIONS FOR NI =  8
NO. OF ITERATIONS FOR FE =  6
```

TABLE 3 – Comparison of simulated and experimental locked cycle
test results for a silver prospect sample

| | | Simulated / Experimental | |
| | | Silver | |
Stream Number	Weight Distribution (%)	Assay (ppm)	Distribution (%)
6	6.4 / 7.0	123 / 128	3.2 / 3.4
9	7.6 / 6.8	391 / 380	11.9 / 9.8
7(T)	98.1 / 98.3	40 / 39	15.6 / 14.4
8(C)	1.9 / 1.7	11.1 / 13.7*	84.4 / 85.6

* Ag assay in ppm x 10^{-3}

added prior to the execution of additional batch tests. Only
after the simulation indicates that satisfactory grade and
recovery would be obtained along with small recycle streams would
a locked cycle test be done to confirm the prediction. An example
of the results of an experimental program conducted along these
lines is shown in Table 3.

The flowsheet consisted of a grinding stage, rougher
flotation, a rougher scavenger and a cleaner. The rougher
scavenger concentrate and the cleaner tails were recycled to the
rougher feed. The average results of the last four cycles of an
eight cycle test are shown in Table 3. Stream number 6 was the
rougher scavenger concentrate, stream 9 was the cleaner tails,
stream 7 the rougher scavenger tails, and stream 8 the final
concentrate. Despite the fact that less than 2% of the feed
weight was floated into the final concentrate there is
quantitative agreement between the simulated results and the
experimental results. This is true not only for the final
concentrate and tails but also for the circulating streams (6&9)
which are always the most difficult to simulate.

The results presented in Figure 3 for a seven cycle locked
cycle test that failed, illustrate that this simulation procedure
is equally capable of predicting bad results. The concentrate
from the first cycle, which is simply a batch test, assayed almost
11% copper. In succeeding cycles with counter current recycle of
the cleaner tails, the concentrate grade decreased progressively
until it appeared to stabilize at the unsatisfactory level of 4 to
5% copper. At the same time the concentrate grade was decreasing
the magnitude of the circulating load was increasing.

The simulation which successfully predicted this behavior was
obviously done after the fact. Had this simulation procedure been
available this locked cycle test would not have been done.

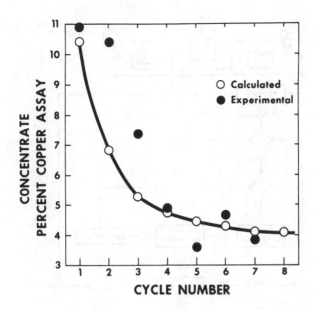

FIGURE 3 – Concentrate assays as a function of the number of cycles for an unsatisfactory locked cycle test

The closest to a full scale test of this procedure which is available [9] was done in an inverse fashion. Inverse because the plant circuit was established and operating then batch bench scale tests were done to reproduce the plant results. The simulation procedure was first used to establish the targets for the batch test when the plant operating results were used as input. From the batch results the continuous circuit was simulated and a locked cycle test was done. The flowsheet for this mill circuit is shown in Figure 4.

The results for all 15 streams are given in the original reference, only the final concentrate and tails are shown in Table 4. The agreement amongst these three data sets is quite acceptable especially when account is taken of the extreme conditions that had to be used in the laboratory flotation machine to emulate the plant performance.

While the foregoing discussion has referred only to flotation results it should be apparent that the simulation procedure applies equally well to any physical separation process. An electrostatic separation of aluminum dross [10] and hydrocyclone classification [11] have both been satisfactorily simulated with this procedure.

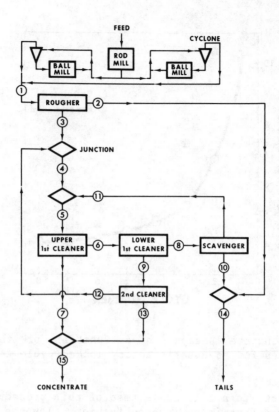

FIGURE 4 - Plant flowsheet

TABLE 4 - Comparison of plant, simulated and laboratory locked cycle test results

	Assays		Distribution (%)			
	% Ni		Weight		Nickel	
	Conc	Tails	Conc	Tails	Conc	Tails
Plant	8.98	0.20	5.6	94.4	72.3	27.7
Simulated	8.76	0.18	6.5	93.5	76.9	23.1
LCT	9.15	0.19	6.2	93.8	76.2	23.9

Rate Equation

The simulation procedure just described treated the separators as black boxes in which a split was made. By recognizing that flotation is a first order rate process the simulation procedure can be put on a firmer base and can be made more flexible. In the following discussion the premise will be made that flotation can be described kinetically with a modified first order rate equation [12, 13, 14]. While there appears to be ample evidence to accept this premise as a reasonable description of the process, any other rate equation could be used equally well.

The modified first order rate equation that has been used is:

$$R = RI[1 - \exp(-k(t+\phi))]$$

where R is the recovery after any time t
RI is the maximum theoretical flotation recovery
t is the cumulative flotation time
k is the first order flotation rate constant (time^{-1})
ϕ is a time correction factor (time)

If data can be fitted to this, or some other rate equation, from batch bench scale data then the recovery to the concentrate, hence the split factor, can be calculated for any desired cumulative flotation time. In the next section it will be demonstrated that the rate equation can be used to optimize the separation time in a given separator.

The collection of data from a batch test in a manner suitable for fitting the parameters in the rate equation requires special care. To minimize differences in performance between operators Roberts et al [15] modified a Denver laboratory flotation cell by fitting it with an angled deflector so that all of the froth would be available for scraping. They also used a froth skimmer that fitted tightly inside the cell width and which skimmed to a constant fixed froth depth. These two cell modifications as well as a cell level controller described by Luttrell and Yoon [16] were employed to maintain a fixed pulp level and thereby a fixed froth depth. Additionally, the air was added through a flowrator with a toggle switch. The agitator speed and air flowrate were selected to produce a modified air flow number of $3x10^{-3}$ [17].

The time correction factor was added in the rate equation because of the difficulty in physically assigning time zero. The clock was always started at the same instant as the air flow and this was nominally time zero. Since it takes a finite time for a stable froth to accumulate on the pulp surface it would seem that a negative correction to the time would be required. Indeed this

TABLE 5 – Replicated batch flotation test results

Cumulative Flotation Time (sec)	Chalcopyrite				Mean	σ	Pentlandite				Mean	σ
			CUMULATIVE RECOVERY									
15	0.491	0.467	0.447	0.406	0.453	0.036	0.338	0.266	0.289	0.189	0.271	0.062
30	0.671	0.628	0.628	0.589	0.629	0.033	0.521	0.411	0.471	0.354	0.439	0.072
60	0.819	0.789	0.800	0.799	0.802	0.013	0.746	0.651	0.732	0.661	0.698	0.048
105	0.899	0.884	0.868	0.883	0.884	0.013	0.876	0.828	0.842	0.846	0.848	0.020
180	0.939	0.934	0.921	0.933	0.932	0.008	0.932	0.916	0.916	0.931	0.924	0.009
300	0.963	0.961	0.945	0.955	0.956	0.008	0.950	0.945	0.939	0.953	0.946	0.007
480	0.977	0.976	0.960	0.968	0.970	0.008	0.958	0.957	0.950	0.963	0.957	0.005
720	0.985	0.985	0.969	0.977	0.979	0.008	0.964	0.964	0.955	0.970	0.963	0.006

Cumulative Flotation Time (sec)	Rock				Mean	σ
15	0.012	0.009	0.010	0.005	0.009	0.003
30	0.017	0.013	0.015	0.010	0.014	0.003
60	0.030	0.024	0.030	0.021	0.026	0.005
105	0.049	0.043	0.042	0.038	0.043	0.005
180	0.073	0.069	0.065	0.065	0.068	0.004
300	0.101	0.093	0.094	0.090	0.095	0.005
480	0.135	0.120	0.123	0.120	0.125	0.007
720	0.186	0.159	0.165	0.159	0.167	0.013

is often observed especially with the least floatable component. On the other hand, the most hydrophobic solids often have some air attached to them during the conditioning period that always precedes the introduction of air which causes them to float faster than they would in the absence of this attached air. Such a phenomonon causes a positive correction factor because flotation appears to have begun before the initiation of air flow. It is not unusual to find a negative correction factor for one mineral and a positive correction factor for another mineral in the same batch flotation tests. Fortunately, with the flotation cell modifications just described the time correction factor is usually small and has little impact on the calculated recovery at the times of greatest interest.

To examine the reproducibility of the data collection procedure just described, Table 5 shows the results of four tests done under nearly the same conditions.

These data were fitted to the first order rate equation with a program that provides a statistically best fit. The program begins with an initial arbitrary estimate of RI then does a least squares fit of the function ln(RI-R)/RI versus the cumulative time. Following this initial fit the initial estimate of RI is decremented by an arbitrary adjustable amount and the least squares fit is repeated. An F ratio statistic is then calculated with the regression sum of squares in the numerator and the residual sum of squares in the denominator. This iterative procedure continues until the F ratio is maximized. Should no

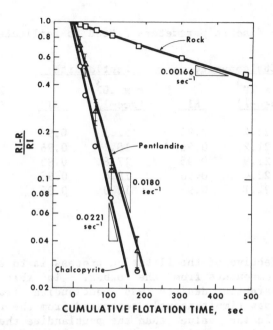

FIGURE 5 – First order rate plots for chalcopyrite, pentlandite and rock

maximum in F be found within the limits of an arbitrary but adjustable value of the correction factor before the value of RI reaches the maximum observed R then the point with the highest value of R is rejected and the routine begun again. The program is allowed to truncate the data only until a minimum of four data points remain but fortunately this seldom occurs.

When the data has been fitted the values of recovery from the fitted equation are printed, the deviation calculated and the fitted parameters are printed.

Figure 5 shows the curves fitted for the average values in Table 5. In the case of pentlandite the range of observed values are indicated, all for an RI of 0.957.

The fitted parameters for the three components, chalcopyrite, pentlandite and rock for each of the four tests as well as for the average results are shown in Table 6.

The resulting equations fitted to the averaged data are shown below.

$$R_{Pn} = 0.957 \, [1 - \exp(-0.018 \, (t+ 3.2))]$$

$$R_{Rk} = 0.235 \, [1 - \exp(-0.0017(t+10.8))]$$

TABLE 6 – Kinetic parameters fitted to replicate test data

Test	Chalcopyrite		Pentlandite		Rock	
	$k \times 10^3$ (sec-1)	RI	$k \times 10^3$ (sec-1)	RI	$k \times 10^3$ (sec-1)	RI
1	23.4	0.96	22.5	0.95	1.5	0.27
2	21.2	0.96	18.1	0.96	1.9	0.21
3	21.9	0.95	17.8	0.95	1.6	0.24
4	22.4	0.96	18.4	0.96	1.8	0.22
AV	22.1	0.96	18.0	0.96	1.7	0.24

Optimization

The objective of the flotation process is to separate one or more of the components from the other. In this instance the primary separation desired was pentlandite from rock. The chalcopyrite was simultaneously separated from the rock but since it was of lessor value than the pentlandite the criterion for optimum separation condtions was based on pentlandite.

The separation efficency [18] has been used for a long time as a technological measure of the efficiency of a physical separation process. It is simply defined as the difference in recovery of the two components to be separated. In this instance

$$S.E. = R_{Pn} - R_{Rk}$$

Differentiating the S.E. and setting the differential equal to zero illustrates that the separation efficiency is at a maximum when the rates of flotation of the two components are equal.

$$\frac{\partial S.E.}{\partial t} = 0$$

$$\left(\frac{\partial R}{\partial t}\right)_{Pn} = \left(\frac{\partial R}{\partial t}\right)_{Rk}$$

$$= (RI)(k) \exp(-k(t + \emptyset))$$

$$(RI)_{Pn}(k_{Pn}) \exp(-k_{Pn}(t + \emptyset_{Pn})) = (RI)_{Rk}(k_{Rk})(\exp-k_{Rk}(t + \emptyset_{Rk}))$$

from which the optimum time, topt, is obtained

$$t_{opt} = \frac{\ln(RI)_{Pn}(k_{Pn})/(RI)_{Rk}(k_{Rk}) - k_{Pn}\emptyset_{Pn} + k_{Rk}\emptyset_{Rk}}{k_{Pn} - k_{Rk}}$$

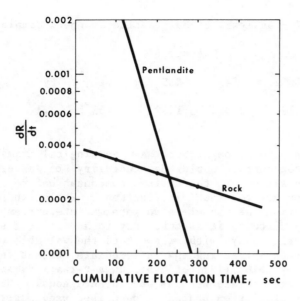

FIGURE 6 - Flotation rate for pentlandite and rock as a function of flotation time

The flotation rates of pentlandite and rock have been plotted in Figure 6 which shows that the rates became equal after 229 seconds of flotation and that is, by definition, the optimum time for this separation stage under this set of chemical and mechanical conditions. By applying this technique sequentially through a flowsheet it is possible to optimize the circuit with respect to flotation times for the physical and chemical conditions selected. This procedure is especially valuable in deciding on the use of primary and scavenger separators because it permits the selection of optimum separations while still maximizing upgrading and recovery. For instance, in the primary separation stage if flotation is carried out for a time shorter than the optimum then a preferably high grade concentrate will be produced with less than optimal recovery. Under some circumstances this may be regarded as preferable but those occasions are rare. On the other hand, continuing the flotation beyond the optimum time will result in unnecessary dilution of the concentrate with gangue; in effect material will be added to the concentrate that is lower in grade than the feed to the separation stage.

Application of this procedure religiously to a high grade feed may well result in a primary separator tailing that is psychologically unsatisfactory because it is too high in assay and from which more desirable component could easily be recovered. A scavenger stage added on this primary separator tailng will solve the problem; the scavenger stage being optimized in the same

TABLE 7 – Analysis of Molybdenite Prospect Sample

Assay (%)

Mo	Cu	Fe	MoS$_2$	Cp	Py	Rk
0.96	0.19	2.08	1.62	0.53	3.1	94.8

manner as the primary separator. A mineralogical examination of the scavenger concentrate would be necessary to ascertain the state of liberation of the desired component before a decision could be made on how to treat it further. While there is no rigorous criterion as to where an intermediate stream should be returned to the circuit, it is customary to add it to a stream of similar chemical assay with respect to the valuable component. Some people advocate treating the intermediate streams separately but a study by NIM [19] of various alternatives showed that recycling the intermediate streams was advantageous. Utilization of the circuit simulation procedure described previously provides the opportunity to examine a host of alternatives, many of which can be rejected without further experimental work.

EXPERIMENTAL EXAMPLE

An experimental test of the foregoing procedures was conducted on a molybdenite prospect [20]. The sample contained molybdenite, chalcopyrite, pyrite and a small amount of tetrahedrite. The main rock mineral was quartz with a small amount of feldspar. The analysis of the feed sample is shown in Table 7.

In the mineral calculation all of the copper was attributed to chalcopyrite then the rock was calculated by difference. The objective of the testwork was to produce a molybdenite concentrate with the minimum level of contamination by copper, iron and rock while maintaining high molybdenite recovery. While no definitive targets were available assays of more than 90% molybdenite (>54.0% Mo) less than 0.2% Cu and 1% Fe were thought to be the least that would be acceptable.

Initial testwork revealed that a combination of a short (low energy input) primary grind followed by flotation of a rougher concentrate with a regrind of the rougher concentrate would produce a high grade concentrate with good recovery. However, because the projected size of the deposit was small a flowsheet that would minimize plant capital cost was desired even though the operating cost might be high. Therefore, tests were undertaken with a long (high energy input) primary grind that would provide sufficient liberation to achieve the target grades.

TABLE 8 – Kinetic results for molybdenite ore flotation

	ROUGHER FLOTATION				1st CLEANER FLOTATION				2nd CLEANER FLOTATION			
	MoS_2	Cp	Py	Rk	MoS_2	Cp	Py	Rk	MoS_2	Cp	Py	Rk
RI	0.95	0.59	1.0	1.0	0.94	0.21	0.25	0.21	0.95	0.51	0.52	0.50
	0.95	0.54	1.0	1.0	0.93	0.11	0.28	0.15	0.93	0.39	0.51	0.43
					0.93	0.28	0.33	0.20				
$kx10^3$	37.5	6.5	0.3	0.24	28	1.4	2.9	1.2	36	8.8	7.8	6.8
(sec^{-1})	38.4	5.8	0.3	0.20	15	3.4	3.1	1.9	51	20.0	25.0	9.0
					21	1.5	3.0	1.8				
θ	5.6	6.0	7.1	-6.1	4.4	4.9	4.5	14.7	4.0	10.4	14.9	11.6
(sec)	2.6	-3.9	11.0	-9.0	3.3	2.4	4.8	-0.7	-0.2	4.9	4.3	9.5
					14.7	14.9	14.5	14.7				

The flotation chemistry required to give good molybdenite flotation while depressng the other sulfides had previously been established through a series of batch tests [21]. The results of these batch tests were analyzed kinetically to maximize the differences in flotation rates between the molybdenite and the gangue sulfides [22]. The minimum energy input to the primary grinding stage was determined from batch tests through a combination of kinetic flotation tests and mineralogical examination of the flotation concentrate. In particular the value of RI for the rock component was especially helpful in identifying the minimum grinding requirement because a locked particle with only a small portion of molybdenite was floatable under the chemical conditions chosen.

Replicate flotation tests were done on the flotation stages in sequence; first a rougher, then a 1st cleaner, then a second cleaner. The optimum separation time was established for the rougher before proceeding to the first cleaner and so on.

The kinetic results from these tests are shown in Table 8. The optimum flotation times obtained from these rate data are shown in Table 9.

The 2nd cleaner concentrate grade at the optimum time averaged 56.4% Mo (94% MoS_2) which satisfied the grade objective; therefore a simulation of a locked cycle test (continuous circuit equivalent) was done with average split factors obtained from all the test data available. For instance two kinetic tests were done

TABLE 9 – Optimum flotation times for molybdenite flotation

Flotation Stage	Optimum Time (sec)
Rougher	133
1st Cleaner	231
2nd Cleaner	69

Table 10 - Experimental split factors for molybdenite flotation

		Mo	Cp	Py	Wt	n
Rougher	SF	0.0564	0.6948	0.9398	0.9445	
	sigma	0.055	0.046	0.011	0.00071	7
1st Cleaner	SF	0.0607	0.935	0.8826	0.687	
	sigma	0.017	0.011	0.023	0.033	5
2nd Cleaner	SF	0.1065	0.71	0.65	0.2227	
	sigma	0.0007	0	0.093	0.0064	2

on the rougher stage then five more tests with the optimum rougher time were done during the cleaner kinetic tests. All of the data used is shown in Table 10.

A locked cycle simulation indicated that a final concentrate assaying more than 59% would be obtained with 94% molybdenum recovery. Since the weight recovery was predicted to be only 1.5% it is obvious that the simulated assays would be extremely sensitive to the weight split factors. To minimize this sensitivity the weight split factors were all calculated from the predicted recoveries of the components. Because the simulated results met all of the objectives with respect to grades, recoveries, recycle flows, and only five iterations were required to reach steady state, a seven cycle locked cycle test was done with these conditions.

The full locked cycle results are shown in Appendix 1. There was good accountability for all components in this test and the weight of the recycle streams stabilized at the fourth cycle values. The concentrate weight was practically constant throughout the test. A comparison of the simulated and experimental locked cycle test results is shown in Table 11.

TABLE 11 - Comparison of simulated and experimental locked cycle test results on molybdenite ore

SIMULATED/EXPERIMENTAL

STREAM	WEIGHT DISTRIBUTION	Mo		Cu		Fe	
		ASSAY	DIST'N	ASSAY	DIST'N	ASSAY	DIST'N
ROUGHER FEED	104.3/105.2	0.99/0.95	107.0/105.0	0.25/0.19	143/106	2.1/2.1	106.1/108.2
ROUGHER CONC	5.8/ 6.8	16.8/13.7	100.8/101.0	1.4 /0.2	43.5/ 7.1	2.3/2.6	6.4/ 8.7
ROUGHER TAILS	98.5/ 98.5	0.06/0.07	6.0/7.5	0.19/0.19	99.1/98.2	2.1/2.1	99.7/102.4
1st CLNR FEED	6.2/ 7.3	17.3/13.6	112.0/108.0	1.4 /0.2	45.6/ 7.9	2.3/2.6	6.9/ 9.5
1st CLNR CONC	1.9/ 2.1	52.0/45.0	105.0/103.0	0.3/ 0.2	3.0/ 1.7	0.9/1.2	0.8/ 1.3
1st CLNR TAILS	4.3/ 5.2	1.5/ 0.9	6.8/5.2	1.9/ 0.2	42.7/ 6.2	3.0/3.2	6.1/ 8.2
2nd CLNR TAILS	0.4/ 0.5	24.9/12.8	11.2/7.2	0.9/ 0.3	2.1/ 0.8	2.5/3.1	0.5/ 0.8
2nd CLNR CONC	1.5/ 1.6	59.8/55.0	94.0/95.9	0.11/0.11	0.9/ 0.9	0.39/0.62	0.3/ 0.5

In general there is excellent agreement between the simulated and experimental results. The minor discrepancy in the molybdenum assay of the final concentrate is due to the extreme sensitivity of the assay to the weight distribution. A similar sensitivity is exhibited by the copper assay in the rougher feed. There is a small difference in the simulated and experimental assays but a large difference in the distribution. It does appear though that the addition of copper sulfide depressant to the rougher stage was on the borderline of being too low. The addition of more depressant in the 1st cleaning stage was effective in depressing virtually all of the copper sulfides so that the experimental distribution of copper to the 1st cleaner concentrate is even less than the simulated value.

Overall the agreement between the simulated and experimental results is sufficient to demonstrate the effectiveness of both the optimization and the simulation procedures.

CONCLUSION

A review has been made of a technique for simulating continuous circuit results from batch data. Also, an effective method of data collection and processing has been demonstrated which permits fitting a first order rate equation to batch test data. With this rate data the optimum time in a given separation stage can be accurately specified. Finally, it was demonstrated that application of these techniques to a real prospect sample permitted the design of an efficient separation circuit. The results of an experimental locked cycle test and the simulated results were in excellent agreement.

REFERENCES

1. Klimpel, R.R., "The Engineering Analysis of Dispersion Effects in Selected Mineral Processing Operations", Fine Particle Processing, ed. P. Somasundaran, AIME, N.Y. (1980).
2. Kitschen, L.P., P.J.D. Lloyd and R. Hartman, "The Centrifugal Mill Experience With a New Grinding System and Its Application", Proc. XIV IMPC CIM, Montreal (1983).
3. King, R.P., "The Use of Simulation in the Design and Modification of Flotation Plants", Flotation, A.M. Gaudin Memorial Volume, AIME, N.Y. (1976).
4. Ford, M.A. and R.P. King, The Simulation of Ore Dressing Plants", Int. Jnl. of Min. Proc. 12, No. 4, 285 (April 1984).
5. Lynch, A.J. et al, "Mineral and Coal Flotation Circuits – Their Simulation and Control" ELSEVIER (1981).

6. Agar, G.E. and W.B. Kipkie, "Predicting Locked Cycle Flotation Test Results From Batch Data", CIM Bulletin 71, No. 799, p. 119, (November 1978).
7. Agar, G.E. et al, "Optimizing the Design of Flotation Circuits" CIM Bulletin 73, No. 824, 173 (December 1980).
8. Agar, G.E., W.B. Kipkie and V.K. Drylie, "Ilmenite Concentration From the Laurentian Titanium Deposit", CIM Bulletin 73, No. 824, 140 (April 1980).
9. Agar, G.E. and R. Stratton-Crawley, "Bench-Scale Simulation of Flotation Plant Performance" CIM Bulletin 75, No. 848, 93 (December 1982).
10. Smith, H.W., "Electrostatic Separation of Aluminum from Dross", CMP Toronto Mtg (September 1982).
11. Stratton-Crawley, R. and G.E. Agar, "Multi-Stage Hydrocyclone Circuit Optimization by Computer Simulation" Proc. 11th Ann. Can. Min. Processors. (1979).
12. Zuniga, H.G., "The Efficiency Obtained by Flotation is an Exponential Function of Time", Bol. Minera, Soc. Nac. Minera, 47, 83 (1935).
13. Morris, T.M., "Measurement and Evaluation of the Rate of Flotation as a Function of Particle Size", Tr AIME 193, 794 (1952).
14. Bushell, C.H.G., "Kinetics of Flotation", TrAIME 225, 226 (1962).
15. Roberts, T., B.A. Frith and S.K. Nicol, "A Modified Laboratory Cell For The Flotation of Coal", Int. Jnl. of Min. Proc. 9, 191 (1982).
16. Luttrell, G.H. and R.H. Yoon, "Automation of a Laboratory Flotation Machine for Improved Performance". Int. Jnl. of Min. Proc. 10, 165 (1983).
17. Harris, C.C., "Flotation Machines", Flotation, A.M. Gaudin Memorial Volume AIME 753, (1976).
18. Schulz, N.F., "Separation Efficiency", Tr AIME 247, 81 (1970).
19. NIM Report.
20. Reesor, R., Senior Thesis, University of Toronto, Dept. Met. and Mat. Sc. Unpublished (1984).
21. Agar, G.E., "Copper Sulfide Depression With Thioglycolate or Trithiocarbonate", to be published in CIM Bulletin (December 1984).
22. Agar, G.E. and J.J. Barrett, "The Use of Flotation Rate Data To Evaluate Reagents", CIM Bulletin 76, No. 851, 157 (March 1983).

APPENDIX 1. : LOCKED CYCLE TEST RESULTS ON MOLYBDENITE ORE

CYCLE	ASSAYS - %			DISTRIBUTION - %			
	Mo	Cu	Fe	WT	Mo	Cu	Fe
1 MO CONC	56.50	.09	.51	1.3	79.6	.6	.3
RO TLS	.060	.20	1.95	93.3	6.1	97.9	90.1
TOTAL	.84	.20	1.93	94.6	85.7	98.5	90.4
*CIR. LOAD	2.44	.05	3.58				
2 MO CONC	55.10	.12	.59	1.6	94.6	1.0	.5
RO TLS	.05	.17	2.03	96.2	5.2	85.9	96.7
TOTAL	.94	.17	2.01	97.8	99.8	86.9	97.2
*CIR. LOAD	1.76	.37	3.30				
3 MO CONC	53.90	.11	.69	1.5	88.7	.9	.5
RO TLS	.06	.19	1.97	99.1	6.4	98.8	96.6
TOTAL	.87	.19	1.95	100.6	95.2	99.7	97.2
*CIR. LOAD	2.54	.41	4.40				
4 MO CONC	55.90	.11	.69	1.5	89.0	.8	.5
RO TLS	.07	.21	1.97	99.4	7.5	109.6	97.0
TOTAL	.88	.21	1.95	100.9	96.6	110.4	97.5
*CIR. LOAD	3.42	.14	5.87				
5 MO CONC	55.60	.13	.60	1.6	95.4	1.1	.5
RO TLS	.07	.19	2.03	99.0	7.5	98.7	99.5
TOTAL	.94	.19	2.01	100.6	103.0	99.8	100.0
*CIR. LOAD	3.30	.16	6.51				
6 MO CONC	55.10	.10	.62	1.7	100.2	.9	.5
RO TLS	.07	.19	2.21	97.7	7.4	97.4	106.9
TOTAL	1.00	.19	2.18	99.4	107.7	98.3	107.4
*CIR. LOAD	1.82	.20	3.41				
7 MO CONC	54.40	.09	.65	1.6	92.2	.7	.5
RO TLS	.07	.19	2.06	98.9	7.5	98.6	100.9
TOTAL	.92	.19	2.04	100.5	99.7	99.3	101.4
2ND C TLS	12.80	.29	3.08	.5	7.2	.8	.8
1ST C TLS	.93	.23	3.21	5.2	5.2	6.2	8.2
*CIR. LOAD	2.01	.24	3.20				
AVG FD/CYCLE	.92	.19	2.02				

DISTRIBUTIONS BASED ON THE AVERAGE FEED PER CYCLE

* = CALCULATED VALUES

SLURRY PIPELINE TRANSPORT OF MINERALS AND COAL

Oner Yucel

Civil Engineering Department and Virginia Center for Coal and Energy Research, Virginia Polytechnic Institute and State University, Blacksburg, Virginia 24061, U.S.A.

ABSTRACT

A review is presented on slurry pipeline transportation of minerals and coal. Primary emphasis is put on past commercial applications, technical characteristics of the various pipeline system components, preliminary costing methodologies, and the hydraulic design criteria for both conventional as well as non-conventional slurries. A number of numerical examples are included for illustration purposes. Although a majority of the technical details pertain to cross-country, large-throughput coal slurry pipelines, appropriate generalizations are also offered for application to other types of slurry pipelines. Also assessed briefly are the overall energy, economic and environmental aspects of slurry pipelines.

KEYWORDS: Slurry transport; slurry pipelines; mineral transport; coal transport; hydraulic design; preliminary costing; conventional slurries; non-conventional slurries; coarse-coal slurries; coal-liquid mixtures.

1. INTRODUCTION

For a number of decades, an exciting expansion of the pipeline industry has emerged in conjunction with the transportation of minerals and coal in slurry form. Today, long-distance, large-throughput pipelines transporting coal, iron, copper, phosphate, limestone and other minerals are a reality. They provide a technically feasible, environmentally superior and economically viable alternative to conventional modes of bulk transportation such as truck, rail or barge. Thompson, et al. [82] define a slurry pipeline as "a two-phase flow pipeline which transports a solid material dispersed in a liquid carrying vehicle." Water is the most widely applied carrying vehicle. However, investigations have shown the viability of a number of other liquids, as well, such as crude oil, methanol (methyl alcohol) and liquid carbon dioxide. Non-water liquids have attracted the interest of designers particularly where water presents problems or from the viewpoint of increased energy transportation. Efficient and economic transportation of energy being the primary objective, various other non-conventional slurries, and especially directly combustible slurries have recently become an important part of research and development in the general area of slurry technology.

1.1 Major Applications of Conventional Slurry Pipelines

In addition to analytical and experimental investigations, significant and varied experience has been accumulated from numerous slurry pipelines in operation or under construction. Fig. 1 and Table 1 highlight the major slurry pipelines around the world. The total length of these major slurry pipelines exceeds 1700 miles (2700 kilometers), with a capacity of transporting over 60 million tons of various products per year, while the cumulative operating experience has exceeded 150 years [5-6,22,28,45,61,64,69,76,82,84-86,89-91].

The first modern study of slurry pipeline technology was undertaken in 1951 by the Consolidation Coal Company in Ohio. A demonstration plant was built, including a 4,000-foot (1300-meter) long horizontal loop and a 1,200-foot (400-meter) long incline loop, both with 12-inch (300-mm) steel pipe, and with its own slurry preparation plant. Systematic investigations were carried out on practically all of the technical aspects of slurry pipelining, including determination of optimum particle size distribution, optimum slurry flow characteristics, power requirements, erosion and corrosion of system components, particle size degradation, shutdown/restart, operational and control procedures, and overall economics. These investigations

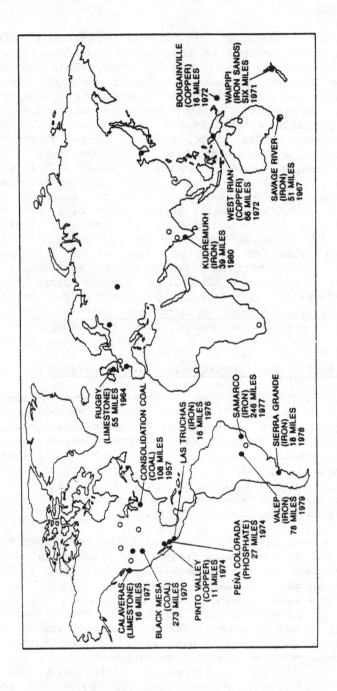

Figure 1 Major Commercial Slurry Pipeline Applications

Table 1 Characteristics of Major Commercial Slurry Pipelines

SLURRY MATERIAL	SYSTEM	COUNTRY	LENGTH MILES	DIAM. INCHES	ANNUAL THROUGHPUT MSPY	INITIAL OPERATION YEAR
COAL	CONSOLIDATION COAL, OHIO	USA	108	10	1.3	1957
	BLACK MESA ARIZONA	USA	273	18	4.8	1970
	BELOVO-NOVOSIBIRSK	USSR	155	17	3.0	1984
LIMESTONE	CALAVERAS CALIFORNIA	USA	17	7	1.5	1971
	RUGBY	ENGLAND	57	10	1.7	1964
	VOEST-ALPINE	TRINIDAD	N/A	N/A	N/A	N/A
	TRINIDAD	TRINIDAD	6	8	0.6	1959
	GLADSTONE	AUSTRALIA	15	8	2.8	1981
COPPER CONCENTRATE	BOUGANVILLE	N. ZEALAND	17	6	1.0	1972
	WEST IRIAN	N. ZEALAND	69	4	0.3	1972
	MURGUL-KBI	TURKEY	40	5	1.0	1973
	PINTO VALLEY ARIZONA	USA	11	4	0.4	1974
IRON CONCENTRATE	SAVAGE RIVER	TASMANIA	53	9	2.3	1967
	WAIPIPI (LAND)	N. ZEALAND	4	8	1.0	1971
	WAIPIPI (SHORE)	N. ZEALAND	1.8	12	1.0	1971
	PENA COLORADA	MEXICO	30	9	1.8	1974
	LAS TRUCHAS	MEXICO	17	10	1.5	1976
	SIERRA GRANDE	ARGENTINA	20	8	2.1	1976
	SAMARCO	BRAZIL	245	20	12.0	1977
	CHONGIN	N. KOREA	61	N/A	4.5	1975
	KUDREMUKH	INDIA	44	18	7.5	1980
	LA PERLA - HERCULES	MEXICO	53 183	8 14	4.5	1982
GILSONITE	AMERICAN - GILSONITE, UTAH	USA	72	6	0.4	1957
COPPER TAILINGS	N/A	JAPAN	44	12	0.6	1968
NICKEL TAILINGS	WESTERN MINING	AUSTRALIA	4.3	4	0.1	1970
PHOSPHATE CONCENTRATE	VALEP	BRAZIL	74	9	2.0	1978
	FOSFERTIL	BRAZIL	9	6	0.9	1981
KAOLIN	FREEPORT GEORGIA	USA	24	10	N/A	1976

lead to the construction of the first long-distance coal-slurry pipeline, which began operation in 1957, known as the "Consolidation Coal Pipeline". The 108-mile (173-kilometers) long pipeline had a capacity of 1.3 million short tons per year. It was operated successfully until June 1963, when it was "mothballed" after having transported some 7 million tons of coal. This was the time when the concept of "unit trains" was introduced by the railroads to compete with the pipeline, and the rail rates were reduced drastically to a level below that of the pipeline [81,89-91].

The next major slurry pipeline in the United States was sponsored by a railroad, the Southern Pacific Transportation Company. The Black Mesa Coal Slurry Pipeline went into operation in 1970. Since then, it has been transporting about 5.5 million tons of coal per year with 99 percent reliability, and is considered to be a milestone in modern pipeline technology. For most of its 273-mile (439-kilometer) length the pipeline is 18 inches (457 mm) in diameter [22,61,91].

Another milestone in slurry pipeline technology is the Savage River pipeline in Tasmania. Since 1967, the 53-mile (85-kilometer) pipeline has also been operated with 99 percent reliability in transporting about 2.3 million tons of magnetite concentrate per year [89]. The 400-mile (640-kilometer) long Samarco pipeline in Brazil, on the other hand, is the longest iron ore pipeline in operation. It went into operation in 1977 and has been transporting 12.0 million tons of hematite type iron ore per year [45]. The Black Mesa, Savage River and West Irian pipelines [89-91] have demonstrated the capability of slurry pipelines to economically overcome various extreme topographic features, which would pose often insurmountable problems for conventional modes of transportation. The Samarco pipeline in Brazil is a convincing example for successfully handling large throughputs of a relatively abrasive ore. The Waipipi pipeline in New Zealand, on the other hand, has the unique feature, that part of the system is submerged in water to connect to an off-shore mooring buoy for direct ship loading.

1.2 Major Applications of Coarse-Coal Slurry Pipelines

Belt conveyors and shuttle cars on rail that are currently employed for the majority of underground and surface haulage of mined coal, are also a primary source of accidents and other hazardous situations. Coarse-coal pipelines with compact and mobile equipment are now recognized as a viable alternative, in this respect, to effectively eliminate these hazards. Investigations carried out since late 1960's have now created a

significantly convincing technology [1,31,41,49,54,57,60,75,83].
A number of pioneering coarse-coal slurry pipeline applications
in the United States, West Germany, Japan and Soviet Union
are referred to as safe, economical and efficient commercial
operations in hauling run-of-mine coal from the mine face to the
surface and to the coal cleaning/preparation plant that may be
located at some distance.

Consolidation Coal Company in the United States has successfully demonstrated the integrated application of underground/
overland coarse-coal pipelining to continuous and long-wall
mining machines as well as conventional coal cleaning facilities. The method, which was patented in 1970, has been applied
to their Loveridge Mine with a performance that has exceeded the
design targets. As a result of over 3.0 million short tons of
cumulative haulage of run-of-mine coal with a 4-inch (102-mm)
top size, an average plant yield of 93 percent has been reported
as of March 1984. The capacity of the system is about 18 short
tons per minute, involving a vertical lift of 900 feet (270
meters) and an overland haulage of 2.4 miles (3.4 kilometers).
The totally computerized programmable logical control system and
continuing research/design improvements are expected to further
increase the system availability [1].

Primarily in association with steep-seam hydraulic mining
operations, both vertical-hoist and overland-type coarse-coal
slurry systems have also been implemented in other parts of the
world. The Gidrougol system in the Soviet Union has two pipes
involving a vertical lift of over 500 meters (1500 feet) [37].
Run-of-mine coals with 90-mm (3.5-inch) and 13-mm (0.5-inch)
top sizes are transported to an overland distance of about 10
kilometers (6.2 miles). The Sunagawa Mine in Japan involves
a series of slurry systems operating since 1965, including a
200-meter (656-foot) hydrohoist segment, handling coal with a
top size of 30 mm (1.2 inches) with a capacity of 100 tons per
hour [81]. At the Hansa Mine in West Germany, mined coal with
a top size of 60 mm (2.4 inches) is lifted 860 meters (2,820
feet) to the surface and hauled overland through two horizontal
pipeline segments of 2 kilometers (1.24 miles) and 3 kilometers
(1.87 miles) in length, with a capacity of about 500 tons per
hour [54].

These and some fifteen other operations with varying,
mostly site-specific system characteristics are now considered
to have demonstrated the safety, technical and economic viability of coarse-coal slurry systems in handling run-of-mine coal
both vertically and horizontally. Consequently, hydraulic
handling of minerals and coal between the underground mine face
and the preparation facility at the surface appears to have even

a greater potential for future implementation, particularly in view of the significant developments that have been accomplished in the computer-driven, integrated control and operation systems.

1.3 Coarse-Coal Ship Loading/Unloading

A parallel development involves hydraulic ship loading/ unloading of minerals and coal. Primarily associated with an offshore, single-point-mooring (SPM) system, coarse-solids slurry pipelining has also been demonstrated to be a technically and economically viable method. A prime example, in this respect, is the IMODCO-MARCONA joint venture at Waipipi, New Zealand, that has been successfully handling iron-sand concentrates since 1971. A number of reports suggest that there are no major technical difficulties that should be involved in coarse-coal slurry applications, identify the various areas that need further to be addressed for the development of optimum design criteria, and illustrate the advantages of coarse-coal ship loading/unloading slurry systems [26,29,52,66,71,77-78].

2. TECHNICAL ASPECTS OF SLURRY PIPELINE SYSTEM COMPONENTS

A slurry pipeline consists mainly of three basic system components: the slurry preparation facility at the point of origin, the pipeline and pumping stations, and the separation or dewatering facility at the receiving terminal. As schematically illustrated in Fig. 2, it may also constitute a major element of total slurry transportation system, which may also include offshore loading/unloading, ocean transport and conventional dry/bulk transportation segments. Today, almost an "off-the-shelf" kind of established technology is available on the various technical design, construction and operation characteristics of each of these system components.

The subsequent sections present a brief assessment of the general technical aspects of the various system components of slurry pipelines. Although the technical details presented pertain mainly to coal slurry pipelines, the general principles and trends are also largely applicable to other types of slurry pipelines that have been implemented in various parts of the world.

2.1 Slurry Pumps

Pumps used to transport solids in current applications fall into the following categories: Reciprocating pumps, centrifugal pumps, hydrohoists, jet pumps and air-lift pumps. For long

295

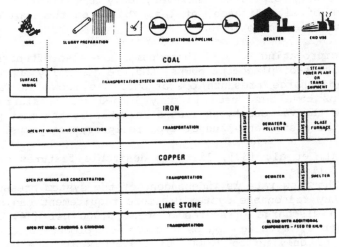

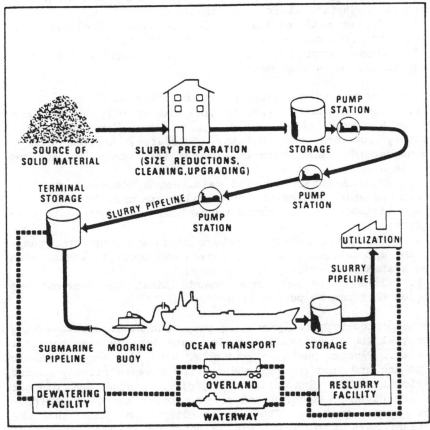

Figure 2   Total Slurry Transport System

distance slurry pipelines, however, only the first two categories are of significance, at least based on the applications
and experience to date.

2.1.1 Reciprocating Pumps: These pumps are also often referred
to as positive-displacement pumps, and are generally appropriate
for high-discharge pressure operations. As shown in Table 2,
these pumps have been used in many long-distance slurry pipelines. Of the various types listed in Table 3, the most popular
ones are the piston and plunger-type reciprocating pumps.

Gandhi, et al. [33], list the desirable features of these
pumps as follows:
1. The flow rate is independent of the system pressure.
2. Any reasonable system pressure requirement can be met.
Pumps with discharge pressures of over 5,000 psi (34,500 kPa)
are available, although pumps capable of producing discharge
pressures of only up to 2,300 psi (15,900 kPa) have been used
to date (in magnetite slurry pipelines).
3. The overall efficiency of the pump, including the
driving mechanism, can be as high as 85 percent.
4. These pumps help determine the pipeline flow rate
without the use of a flow meter.

Their primary disadvantages, on the other hand, are:
1. Their capacity is quite low, namely only up to about
3,900 gallons per minute (0.25 cubic meters per second) at
relatively low pressures. This, of course, implies that a
large number of pumps operating in parallel would be required
to handle high flow rates.
2. Capital investment and maintenance costs, as well as
required operation and maintenance skills are usually high.
3. Variable speed driving mechanisms are required to vary
flow rates.
4. Due to the pulsating nature of flow through the pump,
vibration and fatigue related phenomena may occur in the station
piping system, unless properly designed.
5. Check valve seal requirements limit the size of the
particles that can be pumped in slurry.

The piston-type reciprocating pumps can be of double-acting
design yielding about twice the flow rate for the same physical
pump size. Plunger pumps, on the other hand, are more adaptable
to flushing and lubrication, and thereby have relatively longer
operational life. Schematics of typical piston and plunger
pumps are given in Figs. 3 and 4. Two other types of reciprocating pumps are the GEHO piston-diaphragm pump and the
Mitsubishi Mars pump, which are illustrated in Figs. 5 and 6,

Table 2   Selected Applications of Reciprocating Slurry Pumps

System	Length (km)	Diameter (mm)	Material	Maximum Particle Size	Flow per Pump (m³/hr)	Maximum Discharge Pressure (KPa)*	Pump Drive (kw)	Source of Data
PISTON PUMPS—DOUBLE ACTING, DUPLEX								
CONSOLIDATION	194	250	Coal	8 Mesh	125	8.270	335	Halvorsen (19)
BLACK MESA	490	450	Coal	14 Mesh	475	14.480	1,305	Bechtel, Inc.
CALAVERAS	31	175	Limestone	28 Mesh	190	8.790	520	Thompson et al. (1)
RUGBY	103	250	Limestone	35 Mesh	65	14.480	560	Thompson et al. (1)
TRINIDAD	11	200	Limestone	48 Mesh	70	6.620	135	Kem Keim (20)
COLOMBIA	10	200	Limestone	—	175	1.380	110	Keim (20)
JAPAN	68	300	Slime	—	180	2.930	370	Kawashima (21)
PLUNGER PUMPS								
SAMARCO	396	500	Hematite	100 Mesh	200	14.480	930	Bechtel, Inc.
VALEP	120	225	Phosphate	65 Mesh	132	14.900	670	Bechtel, Inc.
SAVAGE RIVER	95	225	Magnetite	100 Mesh	88	13.790	450	Bechtel, Inc.
SIERRA GRANDE	36	200	Magnetite	100 Mesh	130	9.660	450	Bechtel, Inc.
LAS TRUCHAS	31	250	Magnetite	150 Mesh	190	8.480	520	Bechtel, Inc.
BOUGAINVILLE	31	150	Copper	65 Mesh	110	13.790	520	Bechtel, Inc.
WEST IRIAN	124	100	Copper	100 Mesh	30	15.860	225	Bechtel, Inc.
PINTO VALLEY	20	100	Copper	65 Mesh	45	15.860	260	Bechtel, Inc.
MARS PUMPS								
ENGLISH CHINA CLAY	4	106	Tailings	60 Mesh	74	3.450	112	Sambells (18)
VAAL REEFS, RSA	—	—	Gold Ore Fines	65 Mesh	130	7.860	345	World Mining, Jan 72
OPPU MINING	8	27	Milled Ore	—	75	2.940	100	Mitsubishi
HOKUROKU	72	300	Mill Tailings	65 Mesh	175	4.900	375	Mitsubishi
DIAPHRAGM PUMPS								
CUBAN AMERICAN	—	—	Iron Ore	80 Mesh	—	4.140	—	Keim (20)
N. GERMANY	—	—	Bauxite	3 mm	162	13.180	625	Holthuis, B.V.

*1 KPa = 0.145 psi

Table 3   Application Characteristics of Reciprocating Slurry Pumps

Type of Pump	Maximum Flow m³/hr (gpm)	Maximum Pressure KPa (psi)	Maximum Particle Size (mm)
PISTON	890 (3.900)	15,000 (2,200)	8
PLUNGER	225 (990)	41,000 (6,000)	8
MARS	204 (898)	7,860 (1,149)	1
DIAPHRAGM	325 (1,420)	20,500 (3,000)	4

Table 4   Selected Applications of Centrifugal Slurry Pumps

System	Material	Pipe Length (km)	Flow (m³/hr)	Pipe Dia. (m)	Max. Disch. Pressure (KPa)	No. of Pumps in Series	Source of Data
WAIPPIPI—NEW ZEALAND	Iron Sand	3.3	300	1,465	4,720	6	Bechtel, Inc.
KUDREMUKH—INDIA	Magnetite	67.0	400	1,115	3,100	5	Bechtel, Inc.
SHIRASU—JAPAN	Ash	6.8	600	4,100	2,300	3	

(Note: 1 KPa = 0.145 psi)

298

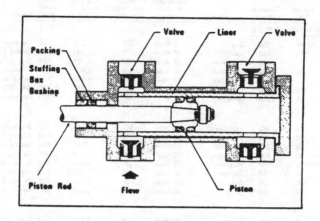

Figure 3   Schematic of a Typical Piston Pump

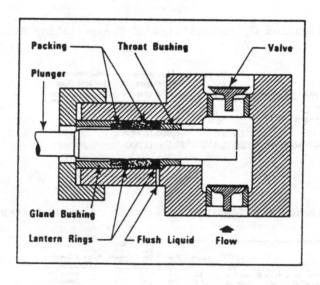

Figure 4   Schematic of a Typical Plunger Pump

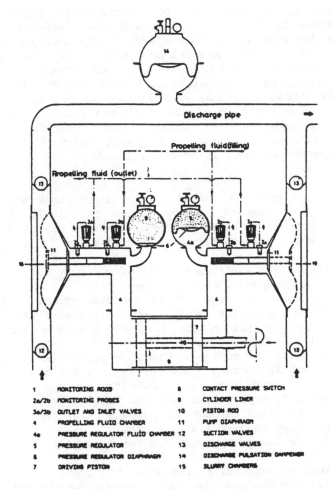

1	MONITORING RODS	8	CONTACT PRESSURE SWITCH
2a/2b	MONITORING PROBES	9	CYLINDER LINER
3a/3b	OUTLET AND INLET VALVES	10	PISTON ROD
4	PROPELLING FLUID CHAMBER	11	PUMP DIAPHRAGM
4a	PRESSURE REGULATOR FLUID CHAMBER	12	SUCTION VALVES
5	PRESSURE REGULATOR	13	DISCHARGE VALVES
6	PRESSURE REGULATOR DIAPHRAGM	14	DISCHARGE PULSATION DAMPENER
7	DRIVING PISTON	15	SLURRY CHAMBERS

Figure 5   Schematic of a Typical Piston-Diaphragm Pump

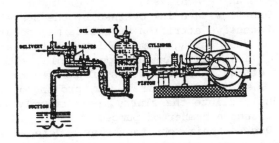

Figure 6   Schematic of a Typical MARS Pump

respectively. With the piston-diaphragm pumps, the major advantage is that the valves and the diaphragm are the only parts to experience major wear. Applications have already demonstrated that special polymers have the potential of significantly diminishing the risk of diaphragm failure [46]. Mars pumps, on the other hand, have successfully been used with highly abrasive solids, such as bauxite, as the valves are the only moving parts in contact with the slurry. The decision as to which pump is more suitable for a specific application is often affected by the size and abrasivity of the solids to be transported as well as by the life of the various parts of the pumps [33].

2.1.2. Centrifugal Pumps: Centrifugal pumps are used extensively for pumping slurry under relatively low discharge pressures, as depicted by Table 4 which exhibits some of the large capacity installation characteristics [33].

The main advantages of centrifugal pumps in slurry applications may be summarized as follows:
1. High flow rates can be achieved with a single unit at a relatively low initial cost. Pumps capable of discharging up to 25,000 gallons per minute (1.6 cubic meters per second) are available.
2. Very few moving and wearing parts are involved.
3. Their operation and maintenance are simple.
4. Maximum size of solids that can be handled is not restricted from practical point of view.
5. The flow through the pump is pulse free.
6. They require little space.
7. No valves are required for operation.

Their drawbacks, on the other hand, are:
1. The maximum discharge pressure is limited to less than 130 psi (900 kPa) for a single-stage pump. With several pumps arranged in series, a maximum practical pressure of 750 psi (5,200 kPa) can be achieved.
2. The flow rate is governed by the system pressure.
3. Excessive attrition of solids may occur due to the high velocity of flow through pump, which may increase the percentage of fines and hence cause appreciable alterations in the hydraulic design characteristics of the slurry.
4. Seal water required for prolonged packing life can significantly dilute the slurry, particularly if the slurry has to pass through a number of pumps.
5. Due to their robustness, as a defense against abrasion, their efficiencies are relatively low.

2.1.3 Hydrohoists, Jet Pumps and Air Lift Pumps: The hydro-hoist consists of one or more chambers that can be filled with solids either in dry form or as a slurry, and then pumped by means of high-pressure water feed into the chamber [11]. A major application of these pumps is in the vertical transportation of coarse solids from deep mines. Their application to long-distance slurry pipelines is not yet popular. Jet and air lift pumps, on the other hand, are suitable only for short distances, and have quite low efficiency.

2.2 Pumping Stations

In each pumping station of a coal slurry pipeline system, as for other slurry pipeline systems, the main components would include a water supply system and related piping, water storage pond, emergency slurry dump pond, agitated or inactive slurry storage tank and reinjection system, slurry pump house, water pump house, cooling tower, instrumentation and operational control room, and a waste treatment facility.

2.3 Line Pipe

Primarily due to the high pressures involved, seamless steel is the only material that has been employed for the line pipes in applications to date for long-distance, conventional-slurry pipelines. In lieu of an existing code, slurry pipelines built to date have been designed using the American National Standard (ANSI) code for Pressure Piping, B31.4, "Liquid Petroleum Transportation Piping Systems", as a guide for safe design [82].

One of the more critical aspects involved in a slurry pipeline design is associated with the prediction of pipe wear due to corrosion/erosion phenomena. For all practical purposes, the problem of corrosion has been adequately researched and understood [36,51]. Wear can become a problem, when partly full (slack) flow conditions occur, usually just downstream of high elevation sections where the hydraulic gradient may intersect the axis of the pipeline. Ensuring a minimum positive pressure for the entire length of the pipeline would prevent such "cavitation" producing conditions. It should also be remarked, in this context, that prototype measurements carried out with the operational slurry pipelines clearly indicate that some of the earlier slurry pipelines might have been over designed with regard to pipe wall thickness.

2.4 Pipeline Construction

Pipeline construction techniques for a coal slurry pipeline are the same as for conventional pipelines. "...The cover from the top of the pipe to ground level would generally be a minimum of 3 feet (about 1 meter), while the actual trench depths would vary with the existing conditions and local regulations..." [25]. A typical right-of-way section is illustrated in Fig. 7.

2.5 Coal Slurry Preparation Facility

Slurry preparation is necessary basically to provide specified slurry characteristics appropriate for hydraulic transportation, and to eliminate the risk of pump and pipe blockage. For this purpose, the raw solid material extracted from a mine source area is processed through a series of mechanical equipment, which crush and grind the size of the particles to suitable values for pipeline transportation, and mixed with specified amount of water to form the slurry. Selection of a suitable preparation system requires a detailed study of the various factors, including the type of material, in particular [8,91].

The hydraulic transportation of solids in water becomes technically easier and requires lower energy when the particles are ground finer in size. However, the separation of solids from water becomes technically more difficult for fine particles and the cost of dewatering process increases. Therefore, an optimum design should take into consideration not only the preparation but also the dewatering components of the whole pipeline system [89-91]. The subsequent discussion deals primarily with preparation of "conventional" coal slurries, while the principles and trends are applicable to other types of slurries as well.

Overall economic considerations for long distance pipeline operations require that the particle top size be kept at a modest level, abrasive wear be minimized, and pipeline steel and pump energy requirements be reduced. For example, the typical size of coal produced at the mine site is 2-in x 0 (5 mm x 0), whereas the final step of burning pulverized coal at the generating plant requires a grind of about 70 percent passing 200 mesh (74 microns). A top size limit of 14-mesh (1.19 mm) is found to be desirable for a long distance pipeline system which is in between the mine product and the boiler requirement. On the other hand, a certain amount of fines, preferably 16 to 23 percent passing 325-mesh (44 microns) for pipelines which are 10 to over 1000 miles (16 to 1600 kilometers) in length, is found

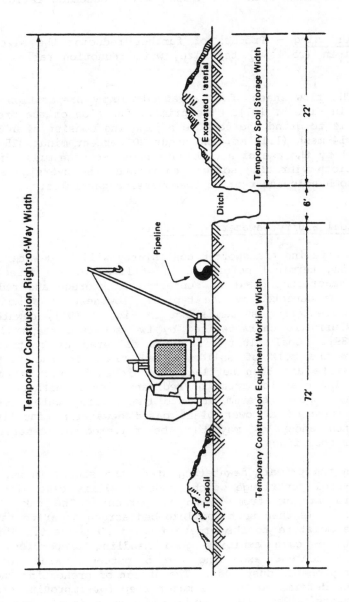

Figure 7 Typical Slurry Pipeline Right-of-Way Cross Section

to be required for providing sufficient slurry viscosity for a
stable pipeline operation [89]. This is usually accomplished in
two major steps:

(a) <u>Crushing</u> - Breakdown of particles into fragments with a top
size less than 1 in. (25.4 mm), with reduction ratios of 4 to
10.

(b) <u>Grinding</u> - Process of further reducing the size smaller
than 1-in (25.4 mm) top size, with reduction ratios of 50 to
100.

The flow sheet of an advanced slurry preparation plant is
shown in Fig. 8. [91]. The primary function of the preparation
plant is to grind the coal to a pipeline consist of 100 percent
minus 14-mesh (1.18 mm) and about 20 percent minus 325-mesh (44
micron) by the use of a combination of two cage mills in series,
wire cloth vibrating screens to recover the oversize and a rod
mill to handle and regrind the oversize particles.

2.6 Coal Slurry Dewatering Facility

A pipeline transported coal slurry will be dewatered at the
receiving terminal before it is utilized either directly in a
power generating plant or subjected to storage and redistribu-
tion.   As depicted by the typical flowsheet of a modern coal
slurry dewatering system shown in Fig. 9 [91], dewatering of
coal slurry is characterized by two separate steps [14,21,56,
61,73,89]. Coal slurry is first filtered or centrifuged to
recover the bulk of solids. Coal fines in the filtrate or
concentrate are then settled and thickened in a sedimentation
unit. The overflow from the sedimentation is either discharged
into a disposal system or into the cooling water system, as
in the case of a power plant based dewatering facility. The
thickened underflow may then be filtered or centrifuged to
recover coal fines.

In the primary dewatering stage the slurry is sent to the
screen-bowl centrifuge which produces slurry cake with surface
moisture varying from 10 to 16 percent. The cake from the
centrifuge is then sent to fluid-bed screen dryer to reduce the
surface moisture to the order of 9 to 10 percent. This final
product of cake exhibits good handling properties, with a
moisture content exceeding the 5 percent value for safety
against probable dustiness. The effect of preheating the slurry
has been determined to be a major step for improving centrifuge
performance. The lower surface moisture from centrifuge was
also found to reduce the subsequent thermal drying. The cake

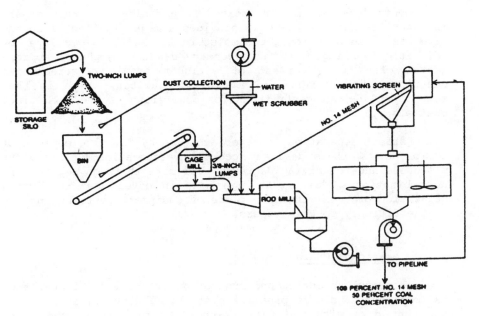

Figure 8   Flowsheet of an Advanced Slurry Preparation Facility

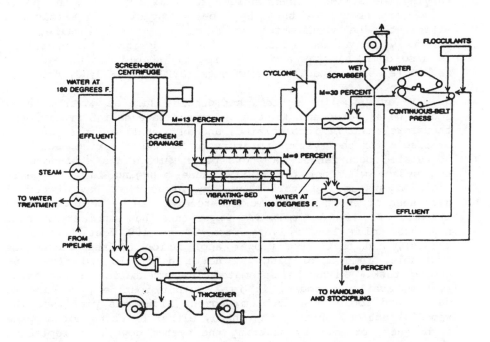

Figure 9   Flowsheet of an Advanced Slurry Dewatring Facility

from centrifuge is further treated in a fluid-bed screen dryer. The sensible heat of the cake provides a major level of evaporative heat needed to lower the moisture in the fluid-bed dryer by adiabatic flashing. In the secondary dewatering phase, approximately 1.5 to 2 percent of effluent from the screen-bowl centrifuge is thickened in a clariflocculator, which is then sent to solid-bowl centrifuges for dewatering of the much finer coal slurry.

Perfect separation is never achieved by any single piece of equipment. Various devices accomplish separation by different processes and in varying degrees. For more thorough separation, a combination of different equipment is often used to form a separation process with each piece of equipment contributing its best efficiency to the process.

2.7 Operational Characteristics

A careful selection and monitoring of the mode of operation is important to slurry pipeline systems. The system is normally not shut down with the full slurry load present in the line. Both start-up and shut-down operations are carried out by varying the solids concentration gradually for a smooth operation. Any amount of throughput below capacity can safely and efficiently be transported in a given slurry pipeline, but obviously not as economically as for the design conditions. It is also obvious, that a slurry pipeline should not be expected to handle above capacity throughputs [5,22,25,38-39,89,98].

In any pipeline system, emergencies such as power outage, should be expected and do occur. Appropriate design provisions, therefore, must be incorporated, and an operational plan must be available for the system operators to implement during each such postulated occurrence. A short power outage and shut-down of the system will not ordinarily cause plugging in a properly designed system. Historically, any plugs that have occurred have been of short length, and therefore, can be deplugged once its location is determined. Experience has indicated, that segmental plug lengths vary from 50 to 100 feet (15 to 30 meters), and that with proper application of the techniques available, removal of plugs is not a difficult and time consuming task. With the automated instrumentation and control systems available today, it is a rather routine procedure to detect and locate the problem segment along the pipeline. The remedial action, on the other hand, either by using hot-tapping techniques, or even by shutting the system down, and replacing the plugged segment of the pipe, seldom takes longer than 48 hours [25, 81-82].

2.9 Communication and Control Systems

The general techniques, principles and equipment involved in the communication and supervisory control systems associated with coal slurry pipelines are not different from those employed for the conventional non-slurry pipelines, such as long-distance oil pipelines. In addition to a telephone service for operations personnel, a microwave communication system would be installed for maintenance bases and associated field units, as well as for transmission of data to the supervisory control system computer at the master control station. At approximately 30-mile (45-kilometer) intervals, microwave repeater antennas would be installed on towers along the pipeline route, ranging in height from 40 to over 300 feet (15 to 100 meters), depending on the topography. For communication with the units located off the primary system, such as mine sites, valve locations and river crossings, narrow band radio installation are usually employed. Each individual system component, such as slurry preparation and dewatering plants and pumping stations would also have its own control room.

A recent advanced-technology application, in this respect, is reported on the completely automated, computer-driven communication/control system at the La Perla-Hercules iron-concentrate pipeline in Mexico [12]. The pipeline system is linked with the two concentration plants and a pelletizer plant through a dedicated microwave communication system, and is controlled from a remote master station. Operations to date have been reported to be virtually trouble-free.

3. HYDRAULICS OF SLURRY FLOWS IN PIPELINES

The overall stability and feasibility of a total slurry transport system depends strongly on the stable hydraulic performance of its pipeline component. As in many engineering processes, none of the components of a system is generally independent from the other components. Therefore, the design and performance characteristics of the pipeline component is also affected strongly by those of the other system components. Clearly, optimum design of the total system may not necessarily correspond to individually optimum design of its various stages or components. Experience has shown that when coal is ground to a size consist of 4 mesh x 0 (4.76 mm x 0), generally optimum characteristics are obtained for a slurry transport system involving a solids concentration of approximately 50 percent by weight [89]. Experiments have also indicated, on the other hand, that a slurry made of finer solids, such as 16 mesh x 0 (1.00 mm x 0) yield lower pipeline transportation costs. However, grinding the coal to a finer consistency would have negative impacts on either end of the pipeline, since both preparation and dewatering costs would be significantly increased. Therefore, it should be remembered, that the hydraulic design of a slurry pipeline must be considered not as an individual component but as a part of the total system.

3.1 Behavior and Classification of Slurry Flows in Pipes

When a slurry is formed with a certain concentration of a specified solids material, and pumped through a pipeline, various flow conditions may be developed. From an engineering viewpoint, these conditions may best be illustrated by examining a typical relationship between the pressure drop (or hydraulic gradient) and the flow rate for slurry flow in a horizontal pipe, sketched in Fig. 10. On the log-log plot, the clear-liquid relationship would be a straight line. If the flow rate (or mean velocity) is sufficiently high, the curve for the slurry would be approximately parallel to the clear-liquid line, indicating somewhat greater pressure drops for the slurry flow. In Fig. 10, this is the region between points 1 and 2, where the solid particles are fully supported by the liquid forming an approximately uniform concentration distribution across the pipe section. If the flow rate is decreased to below point 2, the slurry curve begins to diverge form the clear-liquid line, thus exhibiting increasingly greater pressure drops as compared to those for the clear-liquid. In this range, indicated by the points 2 and 3 in Fig. 10, the solids concentration near the bottom of the pipe increases as the flow rate decreases, thus resulting in a nonuniform concentration distribution. If the flow rate is further decreased to value corresponding 3, the solids along the bottom begin to show the "tendency" of becoming stationary. As the flow rate is further decreased below the value corresponding to point 3, the amount of "stationary" solids also increases, forming a "deposit".

The slurry behavior in the region between points 1 and 2 is typically referred to as the "homogeneous", or more correctly, "pseudo-homogeneous" flow. Once again, in this region, the concentration distribution is roughly uniform, and thus the "gravity" effects are negligible. The region between points 2 and 3, on the other hand, is usually characterized by a non-uniform concentration distribution, and referred to as the "heterogeneous" slurry flow. The velocity at point 3 has been given various names, one of which is "critical deposition velocity". Flows below critical deposition velocity are not of

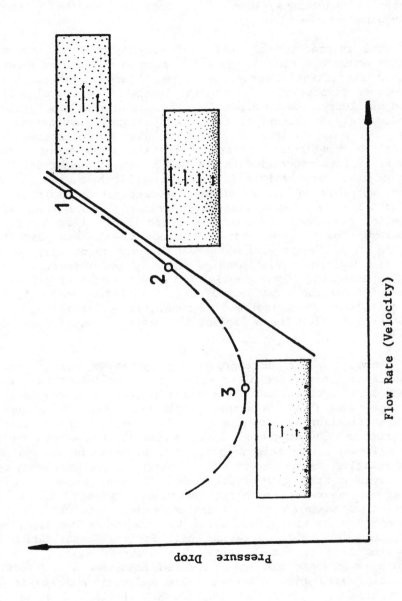

Figure 10   Simplified Features of Slurry Behavior in a Horizontal Pipe

interest, indeed should normally be avoided for design purposes. One obvious reason in this respect, is that transport of solids with partial deposition would be contrary to efficient transportation objectives, even if it does not eventually lead to "plugging" of the pipe.

The interaction of solids of varying physical characteristics with the flow is actually much more complex than the rather simplified description above. Numerous investigations have been devoted to understanding the behavior and classification of slurry flow in pipes. To the scientists, the intriguing fundamental characteristics of solid-liquid suspensions have been the main target. Thus, their ultimate objective is to establish constitutive relations for the slurries of interest and apply them together with the basic laws of conservation. To the engineer involved in design and application, on the other hand, workable equations, whether theoretical or empirical, are essential to establish the design criteria for a given slurry pipeline problem. Behavior of solid-liquid flow may be so complex in many cases, that it has not always been possible to attain a smooth transition between the theory or analytic work and the empirical relationships based on experimental findings or "experience". Consequently, while for some slurry flows the knowledge available may largely be classified as an "art", there is sufficient accumulation of fundamental understanding of some others that allow for a "scientific basis" [5-6,38-39,42,81-82, 87-89].

Among the various approaches proposed, a widely accepted classification divides slurry flows in pipes into three groups. These are homogeneous, hetero-homogeneous (or complex) and heterogeneous flow. A rather simplistic chart exhibiting such a classification is shown in Fig. 11. Numerous theoretical and experimental investigations have suggested, that a more rational classification, in this respect, can be based on the value of the relative concentration C/C_A [87-89]. This parameter is an indicator of the concentration distribution, since C_A is the reference concentration along the pipe axis and C is the concentration measured at a location about (0.08 D) below the top of pipe section of diameter D. Comprehensive experiments conducted with the loop system built for the Consolidation Coal Pipeline lead to the criteria that slurries with C/C_A values of 0.8 or greater can be considered homogeneous. Effects of particle size, pipe diameter, flow velocity and solids concentration on the value of C/C_A are all significant as typified by Fig. 12.

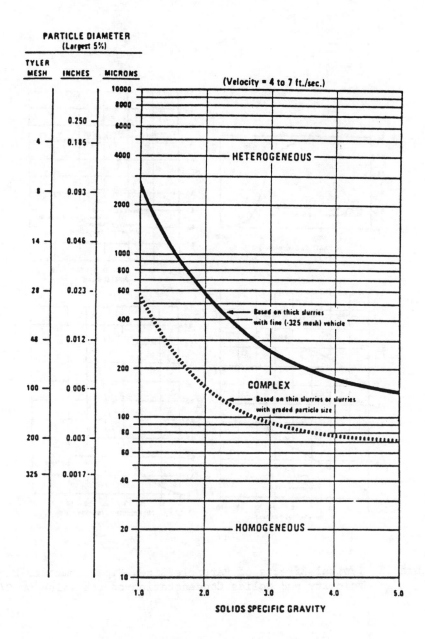

Figure 11   Simplistic Classification of Slurry Behavior in Pipes

312

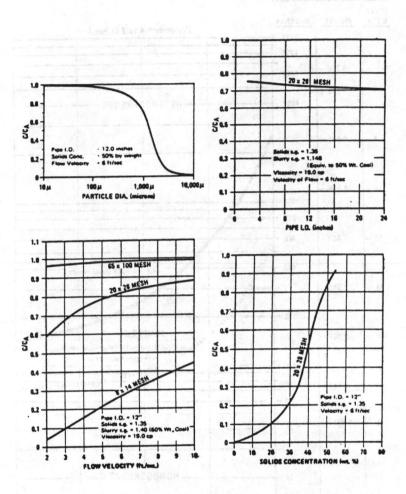

Figure 12 Typical Effects of Particle Size, Pipe Diameter, Flow
Velocity and Solids Concentration on the Value of C/C_A

Previous work has also shown the overall validity of the following relationship [89-91]:

$$\log_{10} (C/C_A) = -1.8 \, w \, / \, (\, \beta k \, V_*)$$
(1)

where, w is the settling velocity of solids, V_* is the shear (or friction) velocity, k is von Karman's velocity distribution coefficient, which has the value of k = 0.04 for clear-liquid flows, and β is a coefficient representing the difference between the diffusivities or motion of solid and liquid particles. Experiments have indicated a 95 percent accuracy for Eq. (1) in predicting the concentration distribution of 14 mesh x 28 mesh (1.19 mm x 0.6 mm) and 20 mesh x 48 mesh (0.84 mm x 0.30 mm) coal-water slurries measured in a 100-mile (161-kilometer) test section along the Consolidation Coal Pipeline. Consequently, Eq. (1) is considered as the basis for classification of slurries in the design of many pipelines.

3.2 Hydraulic Design Criteria for Homogeneous Slurries

Homogeneous slurries can be treated as a new fluid with a unique and generally non-Newtonian rheologic behavior [2,19-20, 42,50,65,72]. Determination of the rheologic properties of slurries is often a formidable task, because of the presence of the solids, however small and uniformly spaced they may be [47,97]. A generalized form of these rheologic models, referred to as the "yield-power-law model", is given by:

$$\tau = \tau_o + k \, s^n$$
(2)

where τ is the shear stress, τ_o is the yield-shear stress, s is the shear-rate, and k, n are constants. One obtains the formulations for a Newtonian fluid with $\tau_o = 0$, k = (μ = viscosity) and n = 1. Similarly, n = 1 yields a Bingham plastic model and $\tau_o = 0$ reduces Eq. (2) to a power law model. The coefficients k and n are also referred to as the coefficient of rigidity or the plastic viscosity (k = η), and flow index, respectively.

A constitutive relation such as Eq. (2) is actually valid only for laminar flow, whereas a universally acceptable formulation does not yet exist for turbulent flow of non-Newtonian fluids. Because homogeneous slurries are supposed to be non-settling, laminar flow would appear to b appropriate regime for implementation to slurry pipeline design. However, all suspensions settle to some extent and at least mildly turbulent conditions would be necessary to eliminate such a risk. Therefore, the transition between laminar and turbulent flow regimes must be determined fairly accurately. Reynolds number would

314

be expected to constitute the predominant parameter, in this respect. However, in view of the general or a simplified version of Eq. (2), various definitions may be applicable for the viscosity, which is the main rheologic variable forming the Reynolds number. Four of these viscosities are [38,40,42]:

(a) The apparent viscosity, μ_a; defined by

$$\mu_a = \tau/s = \tau/(du/dy) \tag{3}$$

where (du/dy) is the velocity gradient;

(b) The effective pipe flow viscosity, μ_e; defined by

$$\mu_e = \tau_w/(8V/D) \tag{4}$$

where τ_w is the wall shear stress, V is the mean flow velocity and D is the pipe diameter;

(c) The viscosity at high shear rates, η_∞; which may be assumed equal to the "coefficient of rigidity" $\eta = k$, for a Bingham plastic fluid.

(d) The Generalized viscosity, k'; which is defined for power law fluids by:

$$k' = k \, 8^{n-1} \, (3n+1/4n)^n \tag{5}$$

where k and n are the coefficients involved in Eq. (2).

With the first three of these viscosities, a conventional Reynolds number can be defined in either of the following forms:

$$R_a = \rho_m VD/\mu_a; \quad R_e = \rho_m VD/\mu_e; \quad R_\eta = \rho_m VD/\eta \tag{6}$$

A generalized Reynolds number has been defined by Metzner (1956) with the use of the generalized viscosity k', as follows:

$$R' = \rho_m D^n V^{n-2}/k' \tag{7}$$

which is actually "not-so-general" [42-43], for it is applicable to power law fluids, only. For Bingham plastic fluids, a better variable for the viscous transition velocity is the Hedstrom number defined by:

$$H = (VD\rho_m/\eta)(\tau_o D/V) = \rho_m D^2 \tau_o/\eta^2 \tag{8}$$

The significance of Hedstrom number is clearly demonstrated in Fig. 13 [42].

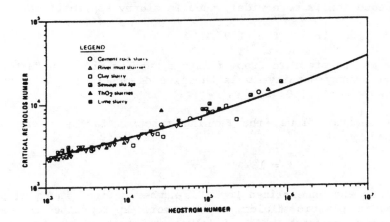

Figure 13   Variation of Reynolds Number with Hedstrom Number

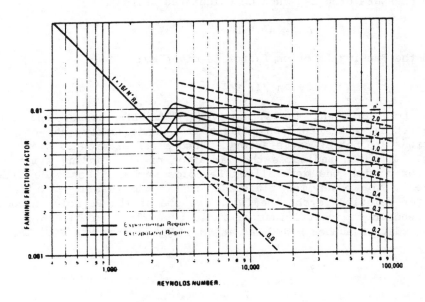

Figure 14   Variation of Fanning Friction Factor with Generalized
Reynolds Number and Flow Index for Power-Law Fluids

The pressure drop for a homogeneous slurry can be calculated, in quite the same manner as applied to single-phase fluids. Rather than the Darcy-Weisbach equation, however, the Fanning equation is more widely used in slurry applications:

$$\Delta p = 2 \, f \, \rho_m \, L \, V^2/D \tag{9}$$

where Δp is the pressure drop, f is the Fanning friction factor, ρ_m is the slurry density, D is the pipe diameter, and V is the mean flow velocity [40,42].

For laminar flows, the Fanning friction factor f can be determined by:

$$f = 16/R' \tag{10}$$

where R' is the generalized Reynolds number defined in Eq. (7). For Bingham plastic fluids, it is more appropriate to use Buckingham's approach which involves integration of Eq. (7) with n = 1, to yield an implicit relationship for the wall shear stress given by [42]:

$$8V\eta/D = \tau_w[1-(4/3)(\tau_o/\tau_w)+(\tau_o/\tau_w)^4/3] \tag{11}$$

The pressure drop can then be calculated using:

$$\Delta p = 4\tau_w L/D \tag{12}$$

and the Fanning friction factor is given by:

$$f = 2\tau_w/(\rho_m V^2) \tag{13}$$

For turbulent flows, however, data are scarce and the theoretical basis is less understood. Consequently, the correlations are not as satisfactory. Such a correlation is shown in Fig. 14, which indicates the variation of the Fanning friction factor f with the generalized Reynolds number R' and the flow index n for power law fluids. For Bingham plastic fluids, a better relation is obtained with the use of the Reynolds number R based on the coefficient of rigidity defined in Eq. (6), and the Hedstrom number H defined in Eq. (8), as shown in Fig. 15.

317

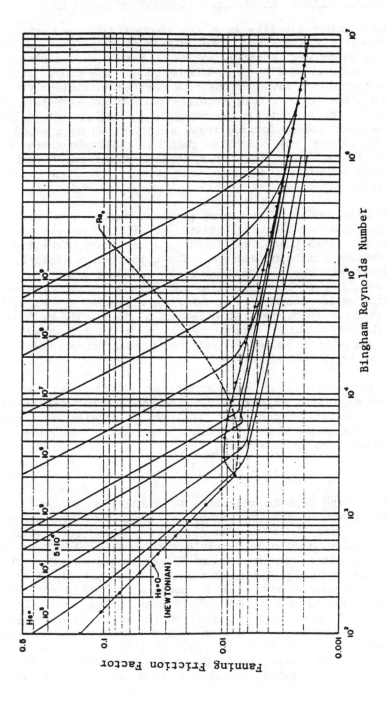

Figure 15   Variation of Fanning Friction Factor with
Bingham Reynolds Number and Hedstrom Number

3.3 Hydraulic Design Criteria for Heterogeneous Slurries

Heterogeneous slurries have two characteristics that make their hydraulic design completely different from that of the homogeneous slurries. Firstly, solids in heterogeneous slurries have a tendency to settle along the bottom of the pipe. Therefore, the turbulence level of the flow must be sufficiently high to prevent deposition. This requires velocities exceeding the so-called "critical deposit velocity." Secondly, the solids and the carrier liquid forming these slurries can be treated as independent phases. This implies that the rheologic characteristics of the carrier liquid can be considered unaffected by the presence of solids, and thus will remain Newtonian in most applications.

Hydraulic design of a slurry pipeline must ensure that no risk of blockage (or clogging) exists, no significant erosion of the pipe wall takes place, and finally, a long-term stability of the slurry flow can be maintained. The objective, therefore, is to transport all the solids completely in suspension, with no solids traveling at low velocities or becoming stationary next to the pipe wall. Consequently, from the viewpoint of hydraulic design, it is perhaps more significant to deal with an "operating velocity", which should be somewhat higher than any critical deposition velocity, yet remain sufficiently low to avoid excessive friction losses.

The pioneering work on deposition velocity was carried out by Durand [23]. It may be safe to state, that all of the subsequent investigations have followed basically the same approach. The basic form of "Durand's parameter" is given in Dimensionless form as follows:

$$F_L = V_D / [2g \ D \ (s_s - 1)]^{1/2} \tag{14}$$

where, V_D is the deposit velocity, D is the pipe diameter, s_s is the dry specific gravity of solids, g is the gravitational acceleration, and F_L is a dimensionless parameter. Various system variables, including particle size concentration, affect F_L, which is often referred to as "Durand's parameter". This variation is actually quite significant for solids particle sizes up to 1 mm, beyond which F_L appears to remain constant, according to Durand's original correlation, shown in Fig. 16. Variation of F_L with concentration, on the other hand, appears to be rather moderate, based on the evaluation of numerous experimental data, as shown in Fig. 17 [89]. Numerous other correlations have also been proposed to incorporate the effects of various other variables, such as particle size,

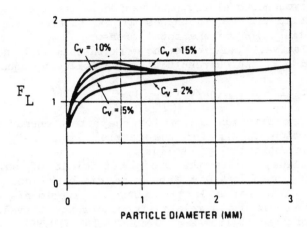

Figure 16   Variation of Durand's Parameter with Particle Size

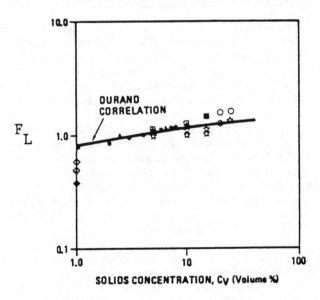

Figure 17   Variation of Durand's Parameter with Concentration

settling velocity, turbulence characteristics and particle size distribution. However, $F_L = 1.0$ appears to be quite representative, if not conservative, a value for most of the data evaluated, particularly considering the subjective nature of observations expected to have been applicable in the determination of the critical deposition velocity [40].

In a heterogeneous slurry, the regime of the flow is always turbulent. Solid and liquid phases may be treated separately, and their contribution to the total pressure drop (or friction loss) is cumulative. The pioneering work in this area was again done by Durand [23], which resulted in the following equation:

$$f_m = f [1 + K C_v [gD/V^2) (s_s-1)/(C_D)^{1/2}]^{3/2} \tag{16}$$

where, f_m is the friction factor for the heterogeneous slurry, f is the friction factor for the carrier liquid, C_v is the volumetric concentration of solids, s_s is the dry specific gravity of solids, C_D is the drag coefficient of solids, V is the velocity of flow, D is the pipe diameter, and K is a coefficient. Data from a large number of experiments carried out to date have yielded a relatively wide range of values for the latter coefficient, from K = 80 to K = 150 [89,90].

3.4 Hydraulic Design Criteria of Hetero-Homogeneous Slurries

Slurries consisting of both fine and coarse size fractions exhibit a complex flow behavior. The relatively coarse solids tend to settle out and hence require higher velocities and turbulence levels to support them in suspension. The finer particles, on the other hand, tend to form homogeneous slurries. The rheology of the carrier liquid is altered by these fine fractions, and a new "vehicle fluid" is formed. The process is quite complex due to primarily three problems. Firstly, it is a formidable task to identify the "rheologically active" fine fractions, and the "rheologically inactive" coarse fractions for a given suspension under specified flow conditions. The second problem is to determine the rheologic behavior of the homogeneous portion of the slurry formed by the fine fractions in the presence of the coarse fractions. Finally, the behavior of the coarse fractions must be determined.

Many investigators have attempted to evaluate the pressure drop due to hetero-homogeneous (or complex) slurries as if they were heterogeneous slurries merely on the basis of a representative particle size of the mixture. A more rational method, recognizing the complexity of these slurries, has been introduced by Wasp, et al. [87-89].

This method has two significant features. One of these is associated with a technique that has been previously used effectively for predicting total bed material load in open channels. The main principle is to divide the particle size distribution into a number of fractions and to account for the relative contribution of each size fraction independently. The total behavior is then determined by simple summation. The second feature of the method is the iterative technique employed to identify the rheologically active fractions of the solids. Initially, all the individual particle size fractions are assumed to behave as "rheologically active" fines forming the "vehicle fluid". The pressure drop calculations for the vehicle fluid are made using the methodology applicable to homogeneous slurries which was discussed earlier. The relative concentration criteria given by Eq. (1) is then used to evaluate the contribution of the homogeneously behaving portion of the solids for each size fraction. The remainder of the solids is assumed to behave heterogeneously, and the relevant pressure drop is calculated using Durand's equation. The "bed portion" of the pressure drop, due to these "rheologically inactive" particles is added to the "vehicle portion" due to the "rheologically active" fraction, to obtain a new total pressure drop. Iterations are continued in this manner until a satisfactory degree of convergence is attained for the total pressure drop, as examplified in Table 5 [47,89,97].

The Wasp method is theoretically sound, and has been proven to yield sufficiently accurate predictions for the pressure drop of hetero-homogeneous slurries. However, this accuracy depends strongly on the proper evaluation of a number of variables [47,97]. Several modifications of Wasp's model have been suggested in an effort to improve on the various problem areas mentioned in the preceding sections. These are primarily related to the determination of the rheologic characteristics of the vehicle portion of the slurry [42].

Another method often used in hydraulic design of complex mixtures is to employ empirical equations derived from experimental data for both the critical deposition velocity and pressure drop. In these equations, the critical velocity and pressure drop are usually expressed as products of the exponential functions of the pipe diameter, representative particle size, specific gravity and concentration of solids, and operating flow velocity. The main drawback of such an approach is, that all the solids are represented by a single particle size, which may not account for the entire slurry behavior.

Table 5 Summary of Calculations with Wasp's Method Applied to a Limestone Slurry

VEL. = 2.30 FPS   PIPE DIAM. = 0.50 FT   SOLIDS CONC. = 40.7% VOL.   SOLIDS SP. GR. = 2.680

ITERATION = 1

SIZE FRACTION TYLER MESH	MEAN DIAMETER (CM)	SETTLING VELOCITY (FT/SEC)	CV SOLIDS (%)	C -- CA	CV VEHICLE (%)	CV BED (%)	SETTLING VELOCITY IN WATER(FT/SEC)	DRAG COEFFICIENT CD	P BED	P VEHICLE (FT-WATER/FT)	P SLURRY
+65	0.0210	0.01890	0.61	0.1927	0.61	0.49	0.13193	2.9	0.0		
65X100	0.0179	0.01381	0.53	0.3003	0.53	0.37	0.09639	4.6	0.0		
100X150	0.0127	0.00691	3.62	0.5476	3.62	1.64	0.04825	12.9	0.0		
150X200	0.0089	0.00343	5.13	0.7415	5.13	1.33	0.02396	36.8	0.0		
-200	0.0042	0.00076	30.81	0.9363	30.81	1.96	0.00528	356.5	0.0		
									0.0	0.00587	0.00587

ITERATION= 2

SIZE FRACTION TYLER MESH	MEAN DIAMETER (CM)	SETTLING VELOCITY (FT/SEC)	CV SOLIDS (%)	C -- CA	CV VEHICLE (%)	CV BED (%)	SETTLING VELOCITY IN WATER(FT/SEC)	DRAG COEFFICIENT CD	P BED	P VEHICLE (FT-WATER/FT)	P SLURRY
+65	0.0210	0.01890	0.61	0.1927	0.12	0.49	0.13193	2.9	0.00683		
65X100	0.0179	0.01516	0.53	0.3003	0.16	0.37	0.09639	4.6	0.00360		
100X150	0.0127	0.00691	3.62	0.5476	1.98	1.64	0.04825	12.9	0.00732		
150X200	0.0089	0.00343	5.13	0.7415	3.80	1.33	0.02396	36.8	0.00269		
-200	0.0042	0.00076	30.81	0.9363	28.85	1.96	0.00528	356.5	0.00073		
									0.02117	0.00560	0.02678

ITERATION= 3

SIZE FRACTION TYLER MESH	MEAN DIAMETER (CM)	SETTLING VELOCITY (FT/SEC)	CV SOLIDS (%)	C -- CA	CV VEHICLE (%)	CV BED (%)	SETTLING VELOCITY IN WATER(FT/SEC)	DRAG COEFFICIENT CD	P BED	P VEHICLE (FT-WATER/FT)	P SLURRY
+65	0.0210	0.02075	0.61	0.1661	0.10	0.51	0.13193	2.9	0.00705		
65X100	0.0179	0.01516	0.53	0.2694	0.14	0.39	0.09639	4.6	0.00376		
100X150	0.0127	0.00759	3.62	0.5186	1.88	1.74	0.04825	12.9	0.00779		
150X200	0.0089	0.00377	5.13	0.7217	3.70	1.43	0.02396	36.8	0.00290		
-200	0.0042	0.00083	30.81	0.9307	28.67	2.13	0.00528	356.5	0.00079		
									0.02229	0.00558	0.02788

ITERATION= 4

SIZE FRACTION TYLER MESH	MEAN DIAMETER (CM)	SETTLING VELOCITY (FT/SEC)	CV SOLIDS (%)	C -- CA	CV VEHICLE (%)	CV BED (%)	SETTLING VELOCITY IN WATER(FT/SEC)	DRAG COEFFICIENT CD	P BED	P VEHICLE (FT-WATER/FT)	P SLURRY
+65	0.0210	0.02088	0.61	0.1644	0.10	0.51	0.13193	2.9	0.00707		
65X100	0.0179	0.01525	0.53	0.2673	0.14	0.39	0.09639	4.6	0.00377		
100X150	0.0127	0.00764	3.62	0.5167	1.87	1.75	0.04825	12.9	0.00782		
150X200	0.0089	0.00379	5.13	0.7204	3.69	1.43	0.02396	36.8	0.00291		
-200	0.0042	0.00084	30.81	0.9303	28.66	2.15	0.00528	356.5	0.00080		
									0.02237	0.00558	0.02795

3.5. Hydraulic Design Criteria for Coarse-Solids Slurries

Primarily because the experience to date is limited to a few, case-specific applications, and the published information is relatively scarce, hydraulic design criteria are not yet established for transportation of slurries involving coarse solids such as run-of-mine coal. The work reported by Trainis [83] is considered important, in this respect. It gives an equation each for the critical deposition velocity and the hydraulic gradient based on the data obtained for horizontal, coarse-coal pipelines in the Soviet Union. Both equations generally follow Durand's concept, but the hydraulic gradient is evaluated for the fine and coarse fractions separately, in a similar manner as in Wasp's model for hetero-homogeneous slurries.

The critical deposition velocity, V_D, is given in metric units as follows:

$$V_D = (gD)^{1/2} [(s_c - s_f) / (k\ C_D\ f\ s_c)]^{1/3} \qquad (20)$$

where D is the pipe size, k is an empirical constant (k = 1.9 for coarse coal), C_D is the drag coefficient for the coarse fraction ($C_D = 0.75$ for coarse coal), f is the Darcy-Weisbach fraction factor for water of the same flow rate as that of the slurry, s_c is the specific gravity of the slurry formed by the coarse fraction, s_f is the specific gravity of the slurry formed by the fine fraction, and g is the gravitational acceleration.

The equation for the hydraulic gradient S_m, of the whole slurry, is also given in metric units as follows:

$$S_m = S\ [1 + (s_s - 1)C_v] + [(gD)^{1/2} (s_s - s_f)C_{vc} / (k\ V\ C_D)] \qquad (21)$$

where S is the hydraulic gradient for water of the same flow rate as that of the slurry, s_s is the dry specific gravity of solids, C_v is the total volumetric concentration, C_{vc} is the volumetric concentration of the coarse fraction, and V is the mean slurry velocity, with the other variables as defined earlier.

Numerous other studies have shown that Durand's concept can be considered applicable essentially to coarse-solids slurry pipelines, as well. For example, Ismail, et al. [49] refer to correlations similar to the one shown in Fig. 18 and comment on the reliability of the criterion of $F_L = 1$, at least for the purposes of a conservative, preliminary evaluation of coarse-coal slurry pipelines. Further investigations are needed,

324

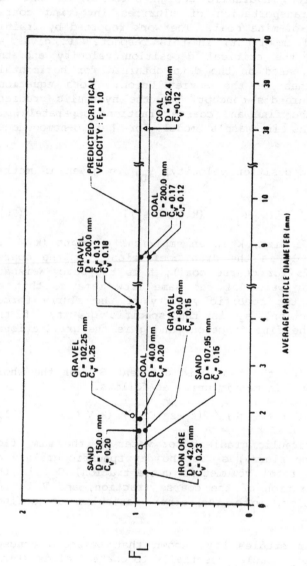

Figure 18 Variation of Durand's Parameter with Coarse-Solids Slurries

however, on such areas as the effect of solids size distribution on the operational stability and determination of optimum representative particle size distribution.

3.6 Hydraulic Design Criteria for Non-Conventional Slurries

Slurries formed by coarse or high-content solids and water and those involving a liquid other than water are usually referred to as the "non-conventional" slurries. The primary basis for such a characterization is not that these are somewhat exotic slurries, but the lack of sufficiently extensive technical information on their behavior, and for that matter, generally acceptable design criteria.

An important class of non-conventional slurries consists of coal-oil, coal-oil-water and coal-water mixtures containing coal concentrations of up to 70-80 percent by weight. Investigations on these slurries have begun fairly recently. It is known by now, however, that they invariably behave as homogeneous mixtures, at least with the help of some "stabilizing" additives, and often have highly complex, non-Newtonian rheology of their own. Their "representative" viscosities are high, and may vary significantly with both time and rate of shear stress applied. Once of the rheologic behavior is determined, then, the general methodology applied to homogeneous slurries discussed earlier would also be applicable to these dense-phase slurries. Other important aspects of hydraulic design for these slurries yet to be fully analyzed are the variations in their rheology due to pumping and performance of the usually "progressive-cavity" type pumps themselves [2,6,24,27,30,55,59, 65,68,72,74].

For slurries involving liquids other than water, such as methanol and liquid carbon dioxide, the main problem again appears to be the shortage of sufficiently extensive information to produce generally acceptable design criteria. A number of studies have shown, however, these non-water-based slurries, too, have a potential to become viable alternatives for transporting minerals and coal, in particular. With appropriate grinding of solids, it appears possible to achieve hydraulic conditions that are similar to those of conventional coal-water slurries. A review of the rheologic, hydraulic and the other technical characteristics of non-conventional slurries studied to date is beyond the present scope.

4. SYSTEM COMPONENT COST ESTIMATING METHODOLOGY

This section presents a summary of the cost functions for estimating the capital and annual operating costs of the various system components for a typical cross-country coal slurry pipeline. The cost functions given are based on an extensive survey and evaluation of the available data and information compiled, concerning not only the existing and planned coal slurry pipelines, but also pipelines transporting other kinds of slurries as well as liquids and gases. The costs presented are with 1982 (end of 1981) prices for the United States, estimated by escalation of the prices of the previous years. Among the more important sources of information on costing slurry pipelines are the reports published by the U.S. Congress Office of Technology Assessment [64], Rieber, et al. [69], Zandi, et al. [99], and Zandi [100], as well as numerous papers published in the proceedings of various international conferences organized by the Slurry Transport Association, and the British Hydromechanics Research Association. Further information was also compiled from the reports as well as through private communications on a number of feasibility studies carried out by various consulting and engineering companies.

4.1 Summary of Cost Estimate Relationships

The cost functions are associated mainly with the following pipeline system components: line pipe, pumping stations, slurry preparation plant, slurry dewatering plant, automated operation, control and communication system and contingencies, all including costs of material, equipment, land, and installation concerning capital costs, and operational and personal costs concerning annual expenses.

After a thorough evaluation of the data available, the relationships employed in the study by Yucel [92] for estimating the capital costs and annual operation and maintenance expenses with 1982 prices applied to coal slurry pipelines in Virginia are summarized below:

Capital Costs ($-millions)

Line Pipe : $LP = (697.019)Dt + (17.311)D + 4.730$

Pumping Stations : $PST = (0.88)T(N_p+1)/N_p + (0.33)T + (0.697)T^{2/3}$

Preparation : $PREP = (4.028)T + (0.756)T^{2/3} + 0.396$

Dewatering : $DEWT = (2.437)T + (5.160)T^{2/3} + 1.000$

Control : $CONT = (0.025)(LP)$

Miscellaneous : $MISC = (0.15)(LP + PST + PREP + DEWT + CONT)$

Total Capital Cost: $PT = LP + PST + PREP + DEWT + CONT + MISC$

Annual Operation and Maintenance Cost ($-millions per year)

Line Pipe : $PLA = (0.050T) + (PW)(NDPY)(VW)$

Pumping Stations : $PSTA = (12)(0.050)(NP)$

Preparation : $PRA = (40)(0.050) + (0.198)T + (0.165)T$

Dewatering : $DWA = (40)(0.050) + (0.220)T + (0.165)T$

Control : $ENA = (PE)(ENY)$

Miscellaneous : $MISA = (0.15)(PLA + PSTA + PRA + DWA + ENA)$

Total O & M Cost : $OMT = PLA + PSTA + PRA + DWA + ENA + MISA$

4.2 Selected Results for a Coal Slurry Pipeline in Virginia

The principles and cost functions described above were applied by Yucel [92] to a preliminary feasibility study carried out for coal slurry pipelines in Virginia. The computer model VIR used for this purpose was an augmented version of the one developed earlier [16-17,96] for a similar study conducted for a magnetite concentrate slurry pipeline in Turkey.

A block diagram of the computer optimization model VIR is shown in Fig. 19. All of the topographic, hydraulic and economic data are retrieved by the subroutine INPUT. The three control subroutines CONT1, CONT2 and CONT3 perform the various functions of managing the program, with respect to the hydraulic and economic calculations and output of the results in both tabulated and plotted formats. The eight route alternatives studies are shown in Fig. 20. The topographic features were typically as shown in Fig. 21 for the 380-mile (608-kilometer) route #1NS. A typical output of hydraulic and cost calculations obtained for the optimum pipeline with an annual throughput of 10 million metric tons per year is given in Table 6, whereas a summary of the results obtained for the route #INS is shown in Table 7.

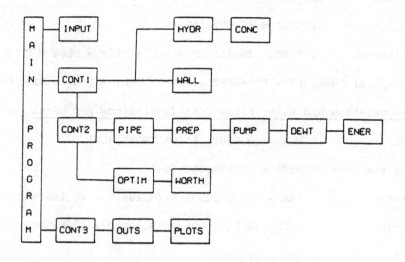

Figure 19   Block Diagram of the Computer Optimization Model VIR

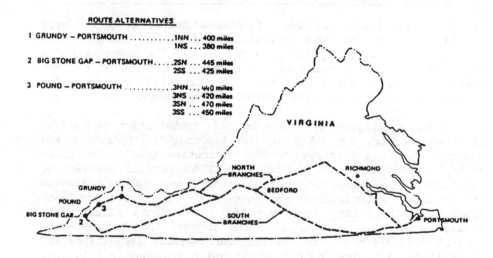

Figure 20   Route Alternatives Investigated for the Conceptual
Coal Slurry Pipelines in Virginia

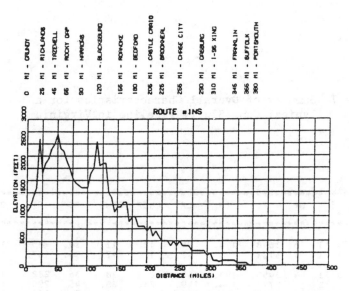

Figure 21 Typical Topographic Profile Characteristics
of Coal Slurry Pipelines in Virginia

Table 6 Summary of Characteristics for a Typical
10-mty Coal Slurry Pipeline in Virginia

```
••••••••••••••••••••••••••••••••••••••••••••••••••••
•    VIRGINIA COAL SLURRY PIPELINE - ROUTE # INS    •
••••••••••••••••••••••••••••••••••••••••••••••••••••

PIPELINE LENGTH         : 380. MILES
OPTIMIZATION CRITERION  : UNIT TRANSPORTATION COST
ANNUAL NET THROUGHPUT   : 10.00 MILLION METRIC TONS/YEAR
ALTERNATIVE SPECIFICS   : 1-LINE PIPELINE WITH FULL LOAD

PIPE SIZE         :  26.76 IN   WGT CONC  :    50.00 %
HYDRAULIC GRADIENT :  0.00477   VOL CONC  :    42.19 %
MIXTURE VELOCITY  :   5.25 FPS  MIX DISCH :  9202.18 GPM

4 PUMP STATIONS AT MI   : 0/ 40/190/295/
NO OF PUMPS PER STATION : 8
AVERAGE HEAD OF PUMPS   : 2156.01 FEET OF WATER
DISCHARGE OF EACH PUMP  : 1314.60 GPM
AVERAGE POWER OF PUMPS  : 1036.64 HP
SUCTION   HEAD OF PUMPS :   32.81 FEET OF WATER
TOTAL POWER REQUIRED    :   21.64 MEGAWATTS
TOTAL ENERGY REQUIRED   :  179.22 MILLION KW-HRS/YR

PIPE WALL THICKNESS AT 5-MI INTERVALS (MILLIMETERS)

19 18 16 14  7 11  9  8 18 17 15 16 16 16 16 16
16 15 15 14 12 10  6  9  8  7 10  9 10  9  8  7  6
 8  7  6  6 18 17 17 16 16 15 15 15 14 13 13 12 12
10 10  9  9  9  8  7  6 16 16 15 15 14 14 13 12 12
11 10 10  9  8  8  7  6  6
----------------------------------------------------
••••••••••••••••••••••••••••••••••••••••••••••••••••
•            SYSTEM COMPONENT COSTS                 •
••••••••••••••••••••••••••••••••••••••••••••••••••••
•                    • CAPITAL  • OPERATION •
•  COST GROUP        • INVEST.  • COST/YR   •
•                    • ($-MILL.)• ($-MILL.) •
•••••••••••••••••••••••••••••••••••••••••••••
• LINE- PE          • 209.269  •   2.785   •
• ENERGY            •   2.500  •   5.377   •
• PUMPING STATIONS  •  65.741  •   2.400   •
• SLURRY PREP       •  44.185  •   5.630   •
• SLURRY DEWT       •  49.321  •   5.850   •
• AUTOMATION        •   5.232  •           •
• MISCELLANEOUS     •  61.831  •   3.582   •
•••••••••••••••••••••••••••••••••••••••••••••
• TOTAL             • 432.683  •  25.348   •
••••••••••••••••••••••••••••••••••••••••••••••••••••
```

Table 7 Summary of Overall Characteristics for a
Typical Coal Slurry Pipeline in Virginia

W	D	S	NP	HP	EN	LP	PREP	DEWT	PS	PT	OM
2.5	13.4	1.13	6	3648.	114.	96.	12.	17.	30.	185.	16.
2.5	13.4	1.13	7	3131.	114.	92.	12.	17.	35.	187.	16.
5.0	18.9	0.73	4	3455.	143.	152.	23.	28.	36.	282.	18.
5.0	18.9	0.73	5	2771.	144.	149.	23.	28.	45.	290.	19.
5.0	18.9	0.73	6	2314.	144.	136.	23.	28.	54.	286.	20.
5.0	18.9	0.73	7	1988.	144.	131.	23.	28.	63.	291.	20.
7.5	23.2	0.57	3	3495.	163.	190.	34.	39.	39.	354.	21.
7.5	23.2	0.57	4	2629.	164.	179.	34.	39.	51.	356.	22.
7.5	23.2	0.57	5	2110.	164.	171.	34.	39.	64.	362.	23.
7.5	23.2	0.57	6	1764.	165.	163.	34.	39.	77.	369.	23.
7.5	23.2	0.57	7	1516.	165.	160.	34.	39.	90.	381.	24.
10.0	26.8	0.48	3	2866.	179.	242.	44.	49.	49.	452.	25.
10.0	26.8	0.48	4	2158.	179.	209.	44.	49.	66.	433.	25.
10.0	26.8	0.48	5	1733.	180.	225.	44.	49.	82.	471.	26.
10.0	26.8	0.48	6	1450.	181.	196.	44.	49.	99.	457.	27.
10.0	26.8	0.48	7	1243.	181.	194.	44.	49.	115.	474.	27.
12.5	29.9	0.41	3	2451.	191.	266.	55.	59.	60.	516.	28.
12.5	29.9	0.41	4	1847.	192.	264.	55.	59.	80.	537.	29.
12.5	29.9	0.41	5	1484.	193.	252.	55.	59.	100.	547.	29.
12.5	29.9	0.41	6	1242.	193.	222.	55.	59.	120.	535.	30.
12.5	29.9	0.41	7	1065.	193.	213.	55.	59.	140.	549.	31.
15.0	32.8	0.37	2	3212.	200.	358.	65.	69.	47.	632.	31.
15.0	32.8	0.37	3	2152.	201.	292.	65.	69.	71.	583.	31.
15.0	32.8	0.37	4	1623.	202.	303.	65.	69.	94.	623.	32.
15.0	32.8	0.37	5	1298.	202.	297.	65.	69.	118.	644.	33.
15.0	32.8	0.37	6	1087.	203.	255.	65.	69.	142.	623.	33.
15.0	32.8	0.37	7	937.	204.	243.	65.	69.	165.	636.	34.
20.0	37.8	0.31	2	2601.	216.	383.	87.	88.	61.	723.	37.
20.0	37.8	0.31	3	1745.	217.	333.	87.	88.	92.	700.	38.
20.0	37.8	0.31	4	1317.	219.	359.	87.	88.	122.	767.	39.
20.0	37.8	0.31	5	1054.	219.	345.	87.	88.	153.	786.	39.
20.0	37.8	0.31	6	883.	220.	293.	87.	88.	184.	762.	40.
20.0	37.8	0.31	7	848.	246.	288.	87.	88.	214.	792.	42.
25.0	42.3	0.27	2	2197.	228.	469.	108.	106.	75.	886.	43.
25.0	42.3	0.27	3	1476.	230.	380.	108.	106.	112.	825.	44.
25.0	42.3	0.27	4	1115.	232.	437.	108.	106.	150.	935.	45.
25.0	42.3	0.27	5	892.	232.	416.	108.	106.	187.	955.	46.
25.0	42.3	0.27	6	749.	233.	397.	108.	106.	225.	977.	46.
25.0	42.3	0.27	7	647.	235.	370.	108.	106.	262.	989.	47.

A detailed discussion of the investigation and the results obtained is presented by Yucel [92,93]. However, two of the interesting features that were observed as common to all alternatives are as follows:

(a) The cost of installed line pipe is by far the major component (over 50 percent) of the capital cost of the total system.

(b) Optimum system cost characteristics do not vary significantly with the number of pumping stations specified. This is quite as expected, because as the number of pumping stations increase, the line pressures decrease, and vice versa. Thus, the sum of line pipe and pumping station costs is somewhat balanced. It should be remarked that this is probably one of the most significant positive features of a slurry pipeline. Actually, it implies that a conceivable range of design variations that may be applied to locations and characteristics of pumping stations in a detailed engineering design study will apparently not alter the overall optimum design characteristics evaluated in a previous feasibility study to any significant degree.

4.3 A Simplified Optimization Methodology

Of the capital cost components referred to earlier, those for the preparation and dewatering facilities and the miscellaneous components are largely direct functions of the annual throughput of the material handled, and therefore, are practically independent of the topography, pipeline route or other variable system characteristics. The variable components of the overall cost relationship, then, are basically those of the line pipe and the pumping stations. In fact, depending on the topography, the number and the individual head/capacity characteristics of the pumping stations directly affect the pipe wall thickness at each along the pipeline. Therefore, the customary procedure is to determine the line pipe and pumping station costs using a segmental calculation technique along a given route. Because the capital cost is invariably the predominant component of long-distance, large-throughput slurry pipelines, minimization of the sum of line pipe and pumping station costs usually yields the optimum system characteristics at least for the purposes of a preliminary feasibility study.

In view of the preceding discussion, a simplified yet quite accurate technique was developed by Yucel [94] for calculating the total system capital cost. The following are the additional observations which formed the basis of the technique:

(a) It can be shown, that for a pipe of given nominal size and material grade, the pipe wall thickness is a direct function of the internal pressure, only.

(b) For a slurry of specified annual throughput, particle size distribution, concentration and operating velocity, the system hydraulic gradient is uniquely determined.

(c) Given a system hydraulic gradient and a topographic profile, the total energy input required by the pumps is constant, and given by the integral of the difference between the energy grade line and the topographic profile.

(d) The total energy input by the pumps directly yields the integral of the line pressure, which determines the integral value of the pipe wall thickness, and for that matter the total weight and the total cost of the line pipe of a given design diameter.

As described in detail by Yucel [94], the preceding considerations lead to the following simplified relationship for the total system capital cost:

$$C_{TS} = C_1 \ (SUM) + C_2 \tag{22}$$

Where C_1 and C_2 are constants for a given annual slurry throughput, whereas SUM is defined by:

$$SUM = S \ L^2/2 - A_t - \sum_{i=1}^{N-1} [L_i \ (\sum_{j=i+1}^{N} H_j)] \tag{23}$$

Clearly, Eq. (23) contains gross quantities only, namely the system hydraulic gradient S, the total length of the pipeline L, the net pumping heads H_j at each of the N pumping stations, and the pumping station spacings L_i, whereas the term A_t is simply the area between the topographic profile and the horizontal plane passing through the lowest pipeline elevation, as indicated in Fig. 22.

Topographic features may sometimes warrant the application of different hydraulic gradients for certain pipeline segments in order to prevent hold-up phenomena. Yucel suggests, that the same concept should lead to a modified form of Eq. (23), again involving only the gross geometric features of hydraulic and topographic profiles, once the locations and pumping heads of the pumping stations are determined, such as by graphical means.

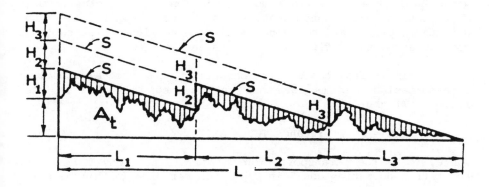

Figure 22   Pumping Heads and Hydraulic Gradients for a
Typical Pipeline on a Given Topography

4.4   Transportation Cost

Disregarding environmental, socio-political and other
applicable factors, the feasibility of a slurry pipeline would
be determined by an overall comparison of the resulting
transportation cost with those applicable to the alternative
modes of transportation that may be available. Such an economic
comparison can be based on the present worth of total costs,
levelized total annual costs or future worth of total costs,
each of which would depend strongly on the economic/financial
scenarios and methodologies applied. These scenarios may vary
considerably not only from one country to another, but also from
year to year and with the economic/financial forecast methods
chosen, a detailed discussion of which is clearly beyond the
present scope. However, the use of levelized annual costs is a
simple, yet rather widely accepted methodology, at least for the
purposes of a preliminary evaluation.

Levelized annual costs consist mainly of the annual operation and maintenance costs and the other expenses, including
financing charges for the capital investment, taxes, insurance,
dividends, and depreciation. These costs must be evaluated
for the project life, for which 30 years is a widely accepted
project basis in today's pipeline applications. A simplified
method would yield:

$$AN = OMT + (R)\ (PT) \tag{24}$$

where AN is the levelized annual cost, OMT is the total annual operation and maintenance cost, PT is the total capital cost and R is the representative fixed annual rate applied to the capital cost to account for financing charges involved.

A simplified basis for comparison of the alternative transportation systems can thus be obtained in either of the following forms:

$$TC = AN \,/\, L; \quad \text{and} \quad TCU = AN \,/\, T \,/\, L \tag{25}$$

where, TC is the "transportation cost" in $ per ton, TCU is the "unit transportation cost" in $ per ton per mile (or kilometer), T is th annual throughput in tons per year, and L is the pipeline length in miles (or kilometers).

The methodology described in the preceding sections deals primarily with the preliminary feasibility costing of a typical long-distance, large-throughput pipeline designed to transport conventional coal-water slurries with approximately 50 percent concentration of coal on a dry weight basis. Pipelines involving other slurries would involve variations regarding these cost functions. The most significant variations that depend on the type of slurry would ordinarily be related to slurry preparation and dewatering systems as well as the operating velocity. A final project design would clearly warrant detailed evaluations of all of the characteristics specific to site and slurry type.

5. BRIEF ASSESSMENT OF ENERGY, ECONOMIC AND ENVIRONMENTAL ASPECTS OF SLURRY PIPELINES

Of the various criteria used to compare various alternative modes of transportation, energy intensity, overall transportation cost, and environmental impact are probably the most significant, if not the deciding ones. While a detailed comparison of slurry pipelines with the other modes of transportation, particularly with the railroads, is not within the present scope, a brief assessment is presented herein.

5.1 Energy Aspects

The energy impacts of slurry pipelines in general, and coal slurry pipelines in particular, have been the subject of a number of investigations, with the following major objectives:
(a) to determine the quantity of the energy expanded by the coal slurry pipelines, and perhaps more significantly,

(b) to establish a common denominator or a representative basis for comparison with the other modes of transportation, particularly with the railroads.

The latter of these items often involves factors which are not necessarily common to different transportation systems and may not even be easily quantifiable. One of the basic measures of "transportation energy" is referred to as the "energy intensity", which is broadly defined as the amount of energy consumed per unit of transportation accomplished. Banks [7] as many investigators, admit that this parameter may often be an inaccurate indicator, since it does not necessarily reflect the impact of the total length, total amount of haulage, and uniformity (continuity) or intermittency of transportation, as well as the availability and even a nonquantifiable factor such as the "kind" or the "value" of the energy that is expanded.

Based on an evaluation of numerous studies, including CBO [15] and Banks [7], Hibbard [44] presented Table 8, and concluded that "...given the uncertainties represented by all of these estimates, the two total... (energy intensity)... values should not be regarded as significantly different... Indeed, it must be argued that when a comparison is made between strictly equivalent options of slurry pipelines and unit trans, it is not possible at this time to distinguish between these options on an energy basis alone... Other criteria such as capital, access, politics, and profitability remain as the major factors which will determine the choice between slurry pipelines and unit trains for moving coal...".

Table 8 Adjusted Energy Intensity Comparison

Categories of Energy Consumption	Coal Slurry Pipeline (Btu/ton-mile)	Unit Train (Btu/ton-mile)
Propulsion	600	370
Vehicles	0	60
Construction	50	100
Maintenance	100	60
Circuitry Factor	(1.1)	(1.51)
Total Energy Intensity	825	890

5.2 Economic Aspects

As a primary motivation behind seeking an alternative to
the current existing systems, overall economic characteristics
of coal slurry pipelines have been the subject of numerous
investigations. The main objective of some of these studies
has been to determine simply as to whether a given coal slurry
pipeline would produce more economical transportation scenarios
than, for example, the railroads. Some studies, on the other
hand, have addressed a more general objective, namely the
overall economic impact of coal slurry pipeline development
upon the related areas of economy, regionally or throughout the
country, or overseas transportation. For more comprehensive and
specific treatment of the related issues, reference is made to
the other studies [4,7,10,13,16,25,29,48,53,64,66,69,92].

The main economic feature of pipelines in general, and coal
slurry pipelines in particular, is that they are predominantly
capital intensive, with up to 70 percent of the costs being
fixed. The reverse is generally the case for railroads, where
up to 80 percent of the costs are variable and affected strongly
by inflation. In an effort to illustrate this point, Aude, et
al. [4] gives the historical and projected future patterns of
slurry pipeline capital cost components in Tables 9 and 10. The
results of similar analyses for the operating cost component,
on the other hand, are given in Tables 11 and 12, respectively.
Aude, et al. [4] also analyzed the projected tariff variations
for both the coal slurry pipeline and rail transportation based
on the records of unit train tarriffs, for the movement of 5 mty
of coal between the same reference points. The results of the
individual tariff projections are given in Figs. 23 and 24,
respectively. Furthermore, Fig. 25 and Table 13 illustrate a
comparison of tariff projections for both modes of transportation. Aude, et al. [4] conclude that the slurry pipeline tariff
in 1983, namely in its first year of operation, is projected to
be 7 percent below the railroad tariff, and this spread will
keep increasing over a 30-year life of coal movement until the
railroad rate is almost three times the pipeline rate. The
increase in spread is attributable to the stabilizing effect of
the comparatively large capital-related component of the pipeline tariff versus the small capital-related component of the
railroad tariff.

Very similar conclusions have been drawn from various other
investigations. It must be remarked again, however, that the
quantitative results above can only be interpreted as indicators
of a general trend, particularly for relatively long-distance
large-throughput haulages. In each case not falling into this
category, the larger the distance and annual throughput, the

Table 9 Historical and Projected Slurry Pipeline Capital Costs

Period	GNPD	Line Pipe[1]	Pipeline[1] Construction	Pump Stations[1] and Other Facilities
1966-1970	4.4	0.4	5.3	2.7
1970-1971	5.0	7.7	10.2	3.7
1971-1972	4.2	1.8	7.9	2.1
1972-1973	5.8	7.0	8.2	2.9
1973-1974	9.6	28.6	11.3	14.9
1974-1975	9.7	39.7	13.6	15.7
1975-1976	5.3	3.2	6.0	5.9
Compound Annual Average Projection[2]	5.7	8.2	7.8	6.3
1976-1983	5.5	7.1	7.0	7.0

[1] Oil & Gas Journal—August 22, 1977
[2] Wholesale Price Index—Machinery & Equipment
[3] Adapted from SHCA

Figure 10 Projected Construction Costs for a Typical 25-mty, 1100-mile Coal Slurry Pipeline

	Pipeline		Pump Station and Other Facilities	Total
	Materials	Construction		
1978-% of Total	28	19	53	100%
1978-Capital-Related Item $/Ton	2.64	1.79	5.00	9.43
Inflation Rates %/Yr. (Table 1)	7.1	7.0	7.0	
1983-Capital-Related Component $/Ton[1]	3.36	2.27	6.34	11.97

[1] Two-Year Construction Schedule

Table 11 Historical and Projected Slurry Pipeline Operating Costs

(ANNUAL PERCENT INCREASE)

Period	Electric Power	Labor	Supplies and Other
1966-1970	1.5[1]	5.5[2]	2.6[3]
1970-1971	6.1	6.6	3.3
1971-1972	7.1	6.6	4.5
1972-1973	7.7	6.0	13.1
1973-1974	12.9	9.5	18.9
1974-1975	21.3	15.0	9.2
1975-1976	13.8	7.9	4.6
1976-1977	9.4	9.8	6.1
Compound Annual Average	7.5	7.5	6.2
Projection			
1977-1980	12.0[4]	7.0	6.1[4]
1980-1985	8.0	6.0	5.5
1985-1990	6.6	6.0	5.5
1990-2010	5.6	6.0	5.5

Table 12 Projected Operating Costs for a Typical 25-mty, 1100-mile Coal Slurry Pipeline

	Power	Labor	Supplies and Other	Total
1978 % of Total	45	15	40	100
1978—Operating Component, $/Ton	1.82	0.61	1.61	4.04
Inflation Rate, %/yr. (Table 3)				
1978-1980	12.0	7.0	6.1	
1980-1983	8.0	6.0	5.5	
1983 Operating Component, $/Ton	2.88	0.83	2.13	5.84

Table 13 Comparison of Projected Tariffs for Slurry Pipelines and Unit Trains

	Transportation Mode	
	Slurry Pipeline	Unit Train
Estimated 1983 tariff	$17.81/ton	$19.08/ton
Annual increase at assumed 5.0% per year general inflation	2.9%/yr.	6.7%/yr.
Projected 2010 tariff	$38.68/ton	$110.72/ton

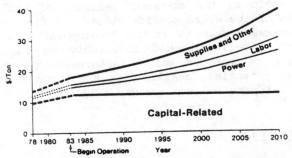

Figure 23 Typical Coal Slurry Pipeline Tariff Projections

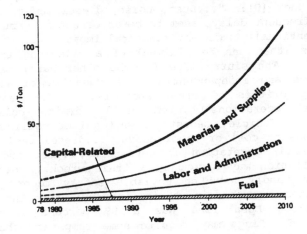

Figure 24 Typical Unit Train Tariff Projections

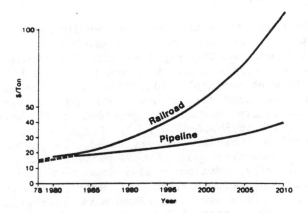

Figure 25 Comparison of Projected Tariffs for
Coal Slurry Pipelines and Unit Trains

more favorable would be the economic features of coal slurry
pipelines in view of the above considerations. A final decision
concerning the economic impacts of the development of a particular coal slurry pipeline should inevitably be based on a
detailed analysis of various other factors, such as market
stability, supply potential, availability and terms of capital,
impact on other shippers and employment.

5.3 Environmental Aspects

To illustrate the environmental superiority of slurry
pipelines particularly over railroads, the following remarks
were made by Buck [9]: "...Dust, noise, diesel fumes, traffic
hazards, and frequent delays seem to be an essential part of the
unit (or conventional) train environmental impact, ... (there is
no) aesthetics (that can be claimed) of a mile-long train of
coal hoppers... The slurry line (on the other hand) is dust-
less, noiseless and independent of the vagaries of weather,
traffic and priorities of other shippers... The best computerized rail operations system still permits collisions
between trains. Vehicle, pedestrian and animal collision are
frequent ...tons of coal or other goods, ...diesel fuel may be
spilled, trackside vegetation is devastated, fires and often
widespread pollution may accompany a train wreck... pipelines
(on the other hand), have an enviable safety record. Coal
slurry pipelines may be in their infancy; but the well-
established Black Mesa line has a 99 percent successful
operation record. Plugs may require some temporary shut-down
of a slurry line, but even if the line must be uncovered and
cut to remove a plug, this is miniscule in time and cost when
compared to a train wreck...".

One of the more frequently discussed environmental aspects
of slurry pipelines is concerned with the quality of water
received at the end of the pipeline and its treatment for
possible reuse as process water or discharge to surface water
courses so as to meet the applicable effluent requirements. In
regards to various constituents and their impact, simulation
studies and laboratory tests have provided quite definitive
indications [9,14,25,56,58,62-63,67,70]. For example, Manahan,
et al. [58] conclude, that "...In many cases, the percentage of
total metal extracted from coal in a 50 percent slurry is very
small... Specifically, environmentally important heavy metals
are not leached appreciably into the water, despite their
significant levels in the coal...". Even more interestingly, it
was further remarked, that "...Because of the rather good
sorption qualities of coal for organic compounds and heavy
metals, serious consideration should be given to the use of coal

slurries for water purification. Thus, impaired water could be used as a source of water for a coal slurry line and could be largely purified by a long-term contact with coal in the slurry..." wastewater were below the detectable limits... (Thus,) water quality problems ... are not severe, and should not impede development of a slurry pipeline system. Remedial measures are available for upgrading the slurry wastewater quality for reuse as process water or to allow discharge to surface watercourses... to clean up the water to whatever level is required...".

In every comparison of slurry pipelines and railroads, the water issue is always assigned as a glaring disadvantage to the slurry pipelines, particularly where the water is known to be scarce. A coal slurry pipeline requires about 800 acre-feet (65 cubic meters) per year for every million ton of coal transported annually. This implies that a pipeline to haul 25 million tons per year requires about 15,000 acre-feet (1,233 cubic meters) of water per year at its source. Compared to the 120,000 acre-feet (10,000 cubic meters) of water required annually for a 7000 megawatts of coal fired electric power plant, this amount of water is certainly not unduly excessive.

Pipeline ruptures and spills are of frequent concern to many. However, based on past experience, frequency as well as incurred damages due to such incidences can now be reduced to a minimum with adequate design considerations, and not necessarily at any substantial costs contrary to the common conceptions. Thus, no long-term impacts are anticipated to the health or safety of any human as a result of a coal slurry spill at any time during the project life.

Historical data on accidents related to slurry pipelines are limited. However, an estimate made based on oil pipeline data by OTA [64] shows, that for coal slurry pipelines to transport about 75 million tons of coal annually, one disability day would result for each 45 million ton-miles (73.6 million ton-kilometers) of coal transported. No estimate on deaths is made due to "lack of supporting data". For railroads carrying the same 75 million tons of coal per year, however, estimates are 1.9 deaths, 304 lost work-day injuries, and 14,000 disability days annually. The study concludes, therefore, that railroads experience much more significant adverse impacts related to occupational health and safety than pipelines. Clearly, it can be argued, that similar statistics may apply to many other parts of the world, as well.

6. CONCLUSIONS

In view of the brief review presented in the preceding sections, the following general conclusions may be offered:

A. The technology of "conventional" slurry pipelines is established, readily available, and practically free from any major and insurmountable problems, from the viewpoints of design, construction, equipment, transport capacity, continuous and efficient operation, line plugging, control of shotdown and restart processes, as well as preparation and dewatering of the slurry involved to any desirable and practical degree without excessive cost, in regards to transporting any mineral or coal, practically to any distance and anywhere in the world. Due to their ability to overcome various topographical difficulties rather easily, it has been and would be possible to economically exploit minerals and coal in many parts of the world, which are otherwise practically inaccessible by other means.

B. While each specific slurry pipeline project will have its own specific characteristics, past experience clearly shows that increasingly favorable conditions for slurry transportation will result with increasing length and annual throughputs. A very significant positive feature of slurry pipelines is that, the fixed capital costs involved invariably constitute by far the dominant portion of the total costs for the entire project life. Therefore, they are much less affected by inflationary factors, which is now a global fact of life, than most other modes of overland transportation. A final decision associated with the development of any slurry pipeline should inevitably be determined by the outcome of a detailed analysis of various specific factors and comparisons with the other modes of transportation that may or may not exist.

C. There is sufficiently convincing evidence and experience that have formed the basis of reliable hydraulic and overall design criteria for the conventional slurry pipelines. The proven techniques of analysis and the indisputable success of the past commercial applications have inevitably lead to recently intensified research and development associated with the "non-conventional" slurries. Relatively more significant applications, in this respect, are offshore ship loading/ unloading pipelines, both underground and overland pipelines handling coarse, run-of-mine coal and minerals, as well as high-concentration coal-liquid mixtures intended for direct combustion in industrial and utility boilers.

7. REFERENCES

Abbreviations : AIChe = American Institute of Chemical Engineers
AIME = American Institute of Mining Engineers
ASCE = American Society of Civil Engineers
BHRA = British Hydromechanics Research Assoc.
EPRI = Electric Power Research Institute
ISCSC = International Symposium on Coal Slurry Combustion
ISCSCT = International Symposium on Coal Slurry Combustion and Technology
ITCOST = International Technical Conference on Slurry Transportation
SME = Society of Mining Engineers
STA = Slurry Transport (Technology) Assoc.
VCCER = Virginia Center for Coal/Energy Research

1. Alexander, D. W., "Loveridge Coarse Coal Slurry Transport System - Performance and Applications," Proc., 8th ITCOST, STA, San Francisco, California, March 15-18, 1983.

2. Alessandrini, A., Kikic, I. and Lapasin, R.,"Rheology of Coal Suspensions," Rheologica Acta, Vol. 22, 1983.

3. Aude, T.C. and Chapman, J.P.,"Coal/Methanol Slurry Pipelines," Proc., 5th ITCOST, STA, Lake Tahoe, Nevada, March 26-28, 1980.

4. Aude, T.C., Wasp, E.J. and Thompson, T.L.,"Coal Transportation Costs/Inflation," Proc., 4th ITCOST, STA, Las Vegas, Nevada, March 28-30, 1979.

5. Bain, A.G. and Bonnington, S.T.,The Hydraulic Transport of Solids by Pipeline, Pergamon Press, London, 1970.

6. Baker, P.J. and Jacobs, B.E.A.,A Guide for Slurry Pipeline Systems, BHRA, Cranfield, England, 1979.

7. Banks, W.F. and Horton, J.F., Efficiency Improvements in Pipeline Systems, Report SSS-R-77-3025-Rev.0, Systems, Science and Software, La Jolla, California, September 1977.

8. Bernstrom, B., "An Evaluation of Grinding and Impact Crushing for Reduction of Coal to Pipeline Size," Proc., 2nd ITCOST, STA, Las Vegas, Nevada, March 2-4, 1977.

9. Buck, A.C., "Negligible Environmental Impact of Coal Slurry Pipelines," Proc., 2nd ITCOST, STA, Las Vegas, Nevada, March 2-4, 1977.

10. Campbell, T.C. and Katell, S., Long-Distance Coal Transport: Unit Trains or Slurry Pipelines, U.S. Bureau of Mines, Information Circular 8690, 1975.

11. Chowdhury, A.K. and Zoborowski, M.E., "High Pressure Coal Slurry Pumping with Hydraulic Exchange Pumps," Proc., 7th ITCOST, STA, Lake Tahoe, Nevada, March 23-26, 1982.

344

12. Chapman, J.P., "Control Systems for Long-Distance Slurry Pipelines," SME-AIME Annual Meeting, Atlanta, Georgia, March 6-10, 1983.

13. Ciliano, R. and Fallah, M.H., Apllicability of Coal Slurry Pipelines to the Appalachian Region, Final Report, Mathtech, Inc., Princeton, New Jersey, September 1978.

14. Cole, R.P., "Fine Pipeline Coal Dewatering Processes, the Current State-of-the-Art and Future," Proc., 9th ITCOST, STA, Lake Tahoe, Nevada, March 21-22, 1984.

15. CBO, Energy Use in Freight Transportation, U.S. Congressional Budget Office, Staff Working Paper, February, 1982.

16. Dabak, T., Technical and Financial Optimization of Slurry Pipelines, Master of Science Thesis, Civil Engineering Department, Middle East Technical University, Ankara, Turkey, May 1979.

17. Dabak, T. and Ger, A.M., User's Manual of the Computer Program OSP, Pipeline Transportation of Solids Series: Report No. H-80-01, Hydraulics Laboratory, Middle East Technical University, Ankara, Turkey, July 1980.

18. Dabak, T., Ger, A.M. and Onder, H., "Experimental Study on Hydraulic Transportation of a Pyrite Ore," Proc., 8th ITCOST, STA, San Francisco, California, March 15-18, 1983.

19. Darby, R., "Rheology of Methacoal Suspensions," Proc., 4th ITCOST, STA, Las Vegas, Nevada, March 28-30, 1979.

20. Darby, R., "Determination and Utilization of Rheological Properties for Prediction of Flow Behavior of Pseudohomogeneous Slurries," Proc., 9th ITCOST, STA, Lake Tahoe, Nevada, March 21-22, 1984.

21. Derammelaere, R.H. and Wasp, E.J., "Mechanical Dewatering, Thermal Drying and Dust Collection," Proc., 8th ITCOST, STA, San Francisco, California, March 15-18, 1983.

22. Dina, M.L., "Operating Experiences at the 1580-MW Coal Slurry Fired Mohave Generating Station," Proc., 1st ITCOST, STA, Columbus, Ohio, February 3-4, 1976.

23. Durand, R., "Basic Relationship of the Transportation of Solids in Pipes - Experimental Research," 5th Congress of the International Association of Hydraulic Research, Minneapolis, Minnesota, 1953.

24. Ekmann, J.M., "Transport and Handling Characteristics of Coal-Water Mixtures," Proc., 4th ISCSC, Orlando, Florida, May 10-12, 1982.

25. ETSI, Environmental Impact Statement on the ETSI Coal Slurry Pipeline, Final Report, Bureau of Land Management, Denver, Colorado, July 1981.

26. Faddick, R.R., "Shiploading Coarse-Coal Slurries," Hydrotransport 8, BHRA, Cranfield, England, 1982.

27. Farris, R.J., "Prediction of Viscosity of Multimodal Suspensions from Unimodal Viscosity Data," Trans., Society of Rheology, Vol. 12, No. 2, 1968.

28. Fister, L.C., Finerty, B.C. and Hill, R.A., "VALEP: The World's First Long-Distance Phosphate Slurry Pipeline," Proc., 4th ITCOST, STA, Las Vegas, Nevada, March 28-30, 1979.

29. Fluor Canada, Ltd., Coal Slurry Pipeline Feasibility Study - 1981, Alberta Economic Development, Energy and Water Resources Departments, Calgary, Alberta, November 1981.

30. Forney, R., "Coal Oil Mixture Firing Direct from Slurry Transported Coal," Proc., 6th ITCOST, STA, Las Vegas, Nevada, March 24-27, 1981.

31. Funk, E.D., "Application of Coarse-Coal Slurries," Proc., 8th ITCOST, STA, San Francisco, California, March 15-18, 1983.

32. Gandhi, R.L. and Aude, T.C., "Slurry Pipeline Design - Special Design Considerations," Proc., Hydrotransport 5, BHRA, Cranfield, England, 1978.

33. Gandhi, R.L., Snoek, P.E. and Carney, J.C., "An Evaluation of Slurry Pumps," Proc., 5th ITCOST, STA, Lake Tahoe, Nevada, March 26-28, 1980.

34. Gandhi, R.L. and Kakka, R.S., "Coal Slurry Pipeline Systems Using Non-Aqueous Mediums," Proc., 8th ITCOST, STA, San Francisco, California, March 15-18, 1983.

35. Ghusn, A.E., "Economics and Design of Slurry Tankers," Proc., 5th ITCOST, STA, Lake Tahoe, Nevada, March 26-28, 1980.

36. Glaeser, W.A. and Hopper, A.T., "Erosion Life in Slurry Pipelines: Possible Sources of Error in Prediction," Proc., 3rd ITCOST, STA, Las Vegas, Nevada, March 29-31, 1978.

37. Gontov, A.E., "Experience of the Coal Hydraulic Transportation Use and Perspectives of Its Development in the Enterprises of the Gidrougol Amalgamation," Proc., 3rd ITCOST, STA, Las Vegas, March 29-31, 1978.

38. Govier, G.W. and Aziz, K., The Flow of Complex Mixtures in Pipes, Van Nostrand Rheinhold Co., New York, 1972.

39. Graf, W.H., Hydraulics of Sediment Transport, McGraw-Hill Book Co., New York, 1971.

40. Graf, W.H., Robinson, M.P., Jr., and Yucel, O., "The Critical Deposit Velocity for Solid-Liquid Mixtures," Proc., Hydrotransport 1, BHRA, Cranfield, England, 1970.

41. Haas, D.B., Husband, W.H.W. and Shook, C.A., "The Development of Hydraulic Transport of Large Sized Coal in Canada," Proc., Hydrotransport 5, BHRA, Cranfield, England, 1978.

42. Hanks, R.W. and Aude, T.C., Slurry Pipeline Hydraulics and Design, Richard W. Hanks Associates, Orem, Utah, and Pipeline Systems Incorporated, Orinda, California, 1982.

43. Hanks, R.W. and Hanks, K.H., "The Importance of Yield Stress in Turbulent Slurry Pipeline Transport of Solids," Proc., 9th ITCOST, STA, Lake Tahoe, Nevada, March 21-22, 1984.

44. Hibbard, W.R., Jr., Comparison of Energy Use Between Unit Trains and Coal Slurry Pipelines, VCCER, Blacksburg, Virginia, April 1982.

346

45. Hill, R.A., Jennings, M.E. and Derammelaere, R.H., "Samarco Iron Ore Slurry Pipeline," Proc., 3rd ITCOST, STA, Las Vegas, Nevada, March 29-31, 1978.

46. Holthuis, C.H. and Simons, P.W.H., "The Economics of Positive Displacement Slurry Pumps," Proc., 6th ITCOST, STA, Las Vegas, Nevada, March 24-27, 1981.

47. Hughes, M.R., Reliability of Bench-Top Viscometer Data in Predicting Hydraulic Transport Characteristics of Conventional Slurries, Master of Science Thesis, Civil Engineering Department, Virginia Polytechnic Institute and State University, Blacksburg, Virginia, March 1984.

48. Hyde, T.E., Economic Comparison Between the Coal Slurry Pipeline Proposed by Houston Natural Gas Corporation and Two Comparable Unit Train Models, Center for Energy Studies, University of Texas at Austin, Austin Texas, January 1979.

49. Ismail. N.M. and Gandhi, R.L., "Prediction of Deposition Velocity for Coarse Solids Slurries," Proc., 8th ITCOST, STA, San Francisco, California, March 15-18, 1983.

50. Ismail, N.M. and Gandhi, R.L., "An Analytical Model for the Prediction of Friction Losses for Homogeneous Slurries," Proc., 9th ITCOST, STA, Lake Tahoe, Nevada, March 21-22, 1984.

51. Jacques, R.B. and Neil, W.R., "Internal Corrosion of Slurry Pipelines, Causes, Control and Economics," Proc., 2nd ITCOST, STA, Las Vegas, Nevada, March 2-4, 1977.

52. Kawashima, T., Tejima, S., and Nagata, K., "Research on Slurry Loading/Unloading of Coal Vessels in Japan," Proc., 1st International SME-AIME Fall Meeting, Honolulu, Hawaii, September 4-9, 1982.

53. Kearney, A.T., Comparison of Projected Maximum Rail and Coal Slurry Pipeline Rates for Transporting Coal to Selected Southeast Utilities, A.T. Kearney, Inc., Report for Coalstream Pipeline Company, Chicago, Illinois, May 1981.

54. Klose, R.B., "The Hydraulic Transport of Coal Suspensions with Coarse Particles," Proc., 6th ITCOST, STA, Las Vegas, Nevada, March 23-26,1982.

55. Lee, H.M., "An Overview of Proposed Coal Slurry Technologies and Their Cost-Saving Applications," Proc., 7th ITCOST, STA, Lake Tahoe, Nevada, March 23-26, 1982.

56. Leininger, D. Erdmann, W. and Kohling, R., "Dewatering of Hydraulically Delivered Coal," Proc., Hydrotransport 5, BHRA, Cranfield, England, 1978.

57. Mlasbury, R.G., Kermit, P. and Saad, A., "A Jet Pump Injector for Coarse Coal Face Hydraulic Haulage," Proc., 4th ITCOST, STA, Las Vegas, Nevada, March 28-30, 1979.

58. Manahan, S.E., Godvin, J.A. and Shinn, M.H., Quality and Traetment of Water Involved with Coal Slurry Transport by Pipeline, Missouri Water Resources Research Center, Columbia, Missouri, March 1980.

59. Meyer, E.G., "A Novel Coal-Based Slurry System," Proc., 8th ITCOST, STA, San Francisco, California, March 15-18, 1983.

60. Miura, H. and Mase, S., "Operation and Maintenance of Slurry Transportation System at Hydraulic Mine," Proc., 4th ITCOST, STA, Las Vegas, Nevada, March 28-30, 1979.

61. Montfort, J.G., "Operating Experience of the Black Mesa Pipeline," Proc., 5th ITCOST, STA, Lake Tahoe, Nevada, March 26-28, 1980.

62. Moore, J.W., "Water Quality and Slurry Pipelining - A Summary of the Literature," Proc., 7th ITCOST, STA, Lake Tahoe, Nevada, March 23-26, 1982.

63. Moore, J.W., "Removal of Heavy Metals from Slurry Wastewater by Ion Exchange and Precipitation," Proc., 9th ITCOST, STA, Lake Tahoe, Nevada, March 21-22, 1984.

64. OTA, A Technology Assessment of Coal Slurry Pipelines, U.S. Congress Office of Technology Assessment, Report No. OTA-E-60, Washington, D.C., March 1978.

65. Onischak, M., Gupta, K.K., Babu, S.P. and Pouska, G.A., "Rheology and Transportation Characteristics of a Coal Gasification Feed Slurry," Proc., 5th ITCOST, STA, Lake Tahoe, Nevada, March, 26-28, 1980.

66. Orr, L.P. and Stein, J.A., "Staten Island Coal Export Terminal, Summary of Feasibility Studies," Proc., 8th ITCOST, STA, San Francisco, California, March 15-18, 1983.

67. Peavy, H.S., "Water Quality as a Factor in the Reuse or Disposal of Coal Slurry Transport Water," Proc., 7th ITCOST, STA, Lake Tahoe, Nevada, March 23-26, 1982.

68. Priggen, K.S., Scheffee, R.S. and McHale, E.T., "Pipelining of High-Density Coal-Water Slurry," Proc., 9th ITCOST, STA, Lake Tahoe, Nevada, March 21-22, 1984.

69. Rieber, M. and Soo, S.L., Coal Slurry Pipelines: A Review and Analysis of Proposals, Projects and Literature, EPRI Research Project RP 1219, Palo Alto, California, August, 1982.

70. Rogozen, M.B. and Margler, L.W., "Environmental Impacts of Coal Slurry Pipelines and Unit Trains," Proc., 3rd ITCOST, STA, Las Vegas, Nevada, March 29-31, 1978.

71. Roseman, D.P. and McDonough, W.D., "Technical Options in the Marine Transportation of Coal Slurry," Proc. 8th ITCOST, STA, San Francisco, March 15-18, 1983.

72. Round, G.F., "Hydraulic Transport of Solids in Pipelines," Applied Mechanics Reviews, 35(10), October 1982.

73. Sandhu, A.S., "Water Quality Simulation for the Design of ETSI Slurry Wastewater Treatment Facilities," Proc., 7th ITCOST, STA, Lake Tahoe, Nevada, March 23-26, 1982.

74. Santhanam, C.J., Dale, S.E., Peirson, J.F., Burke, W.J. and Hanks, R.W., "Coal-Liquid Carbon Dioxide Slurry Pipeline Technology," Proc., 9th ITCOST, STA, Lake Tahoe, Nevada, March 21-22, 1984.

348

75. Sims, W.N., "Coarse Coal Handling by Marconaflo Handling Systems," Proc., 8th ITCOST, STA, San Francisco, March 15-18, 1983.

76. Stephan, F. and Navrade, D.H., "Experiences During Commissioning of the Iron Concentrate Pipeline at O.E.M.K./ U.S.S.R.," Proc., 9th ITCOST, STA, Lake Tahoe, Nevada, March 21-22, 1984.

77. Sterry, W.M., "Pacific Bulk Transport System," Proc., 7th ITCOST, STA, Lake Tahoe, Nevada, March 23-26, 1982.

78. Sterry, W.M., Yucel, O. and Roseman, D.P., "Atlantic Coal Transportation System (ACTS)," Proc., 8th ITCOST, STA, San Francisco, California, March 15-18, 1983.

79. Stripling, T.E., "Hydraulic Design of Large Diameter, Long Distance Coal Slurry Pipelines," Proc., 4th ITCOST, STA, Las Vegas, Nevada, March 28-30, 1979.

80. Thompson, T.L., Frey, R.J., Cowper, N.T. and Wasp, E.J., "Slurry Pumps, A Survey," Proc., Hydrotransport 2, BHRA, Cranfield, England, 1972.

81. Thompson, T.L. and Aude, T.C., "Slurry Pipeline Design and Operation Pittfalls to Avoid," Joint ASME Mechanical Engineering and Pressure Vessels and Piping Conference, Mexico City, Mexico, September, 1976.

82. Thompson, T.L. and Aude, T.C., "Slurry Pipelines, Design, Research and Experience," Journal of Pipelines, Vol. 1, 1981.

83. Trainis, V.V., Parameters and Regimes for the Hydraulic Transportation of Coal by Pipelines, Bureau of Mines and National Science Foundation Joint Report No. TT-72 51038, Washington, D.C., 1977.

84. Venton, P.B. and Cowper, N.T., "Slurry Pipelines in the Cement Industry," Proc., 3rd ITCOST, STA, Las Vegas, Nevada, March 29-31, 1978.

85. Venton, P.B., Cunningham, G. and Tennant, J.D., "Bauxite Transport by Slurry Pipeline," Proc., 4th ITCOST, STA, Las Vegas, Nevada, March 28-3-, 1979.

86. Venton, P.B., "The Gladstone Limestone Pipeline," Proc., Hydrotransport 8, BHRA, Cranfield, England, 1982.

87. Wasp, E.J., Regan, T.J., Withers, J., Cook, P.A.C. and Clancy, J.T., Cross-Country Coal Pipeline Hydraulics," Pipe Line News, July 1963.

88. Wasp, E.J., Aude, T.C., Kenny, J.P., Seiter, R.H. and Jacques, R.B., "Deposition Velocities, Transition Velocities and Spatial Distribution of Solids in Slurry Pipelines," Proc., Hydrotransport 1, BHRA, Cranfield, England, 1970.

89. Wasp, E.J., Kenny, J.P. and Gandhi, R.L., Solid-Liquid Flow: Slurry Pipeline Transportation, Trans Tech Publications, Clausthal, West Germany, 1977.

90. Wasp, E.J., "Coal Slurry Pipelines for the Next Decade," Mechanical Engineering, December 1979.

91. Wasp, E.J., "Slurry Pipelines," Scientific American, Vol. 249, No. 5, November 1983.

92. Yucel, O., Coal Slurry Pipelines in Virginia: A Preliminary Feasibility Study, Final Report, VCCER, Virginia Polytechnic Institute and State University, Blacksburg, Virginia, June 1982.

93. Yucel, O., Atlantic Coal Transportation System: Coal Slurry Pipeline, Final Report, VCCER, Virginia Polytechnic Institute and State University, Blacksburg, Virginia, June 1982.

94. Yucel, O., "Simplified Optimum-Cost Design of a Slurry Pipeline Along a Given Route," Journal of Pipelines, Vol. 4, 1984.

95. Yucel, O., "Potential Feasibility of Integrated Coal-Water Mixture Transportation Systems," Proc., 6th ISCSCT, Orlando, Florida, June 25-27, 1984.

96. Yucel, O., Ger, A.M. and Dabak, T., "Computer Application of a Preliminanry Feasibility Model to a Turkish Iron Ore and Virginia Coal Slurry Pipeline," Proc., 7th ITCOST, STA, Lake Tahoe, Nevada, March 23-26, 1982.

97. Yucel, O. and Hughes, M.R., "Sensitivity of Pressure Drop to Particle Size Distribution and Related Rheologic Characteristics of Hetero-Homogeneous Slurries," Proc., 9th ITCOST, STA, Lake Tahoe, Nevada, March 21-22, 1984.

98. Zandi, I., Advances in Solid-Liquid Flow in Pipes and Its Application, (Ed.), Pergamon Press, New York, 1971.

99. Zandi, I. and Gimm, K.K., Transport of Solid Commodities via Freight Pipeline: Freight Pipeline Technology, Vol. 2, Final Report, University of Pennsylvania, Report No. DOT-TST-76T-36, Philadelphia, Pennsylvania, July 1976.

100. Zandi, I., "Coal Slurry Pipeline Cost Model," Journal of Pipelines, Vol. 3, 1982.

LIST OF PARTICIPANTS

Agar, G.E.*
Mineral Processing Research
INCO, Ltd. 2060 Flavelle
   Blvd.
Sheridan Park, Mississauga,
Ont. L5K 1Z9 CANADA

Atalay, U.
Mining Engineering Dept.,
Middle East Technical Univ.
Ankara, TURKEY

Ayok, T.
Kimya Muhendisligi
Arastirma Bolumu TUBITAK Marmara
   Enst.
PK. 21 Gebze, Kocaeli, TURKEY

Barhana, S.
Kutahya 100.Yil Gumus Mademi
Isl. Muessesesi Etibank PK. 106
Kutahya, TURKEY

Bayraktar, I.
M.T.A., Teknoloji Dairesi
Cevher Zenginlestirme Servisi
Ankara, TURKEY

Berk, H.
Mining Engineer, Izmir Caddesi
Koc is Hani No: 25/7 Kizilay
Ankara, TURKEY

Bruynesteyn*
A. Bruynesteyn and Associates
2175 Greylynn Crescent North
Vancourver B.C. CANADA V7J 2X6

Bulutcu, N.
I.T.U. Kimya Metallurji Fakultesi
Kimya Muh. bolumu Ayazaga Kampusu
Istanbul, TURKEY

Burt, R.O.*
Tantalum Mining
Corporation of Canada Ltd.,
Box 2000, Lac du Bonnet Manitoba
ROE LAO CANADA

Cakir, M.
Etibank, Bursa Maden Isletmeleri
Muessese Muduru, Cekirge
Bursa, TURKEY

Carvalho, J.M.R.
Grupo Extracco Liquido-Liquido
Seccao 41, I.S.T. Av-Rovisco
Pais 1096 Lisbao Codex, PORTUGAL

Casquinha, H.
Laboratorio Nacional De Engenharia
E Tecnologia Industrial, Estrada
do Paco do Lumiar, 1699 Lisboa
Codex, PORTUGAL

Dabak, T.
Virginia Center for Coal and
   Energy
Research, Virginia Polytechnic
   Inst.
and State University,
Blacksburg, VA 24061, U.S.A.

De Cuyper, J.*
Laboratory of Mineral Treatment,
Unversite Catholique
de Louvain B. 1348 Louvan-La Neuve
BELGIUM

Dias, S.
I.S.T. Avenue Rovisco Pais 1096
Lisbao Codex, PORTUGAL

Dogan, Z.M.+
Dept. of Mining
METU, Ankora, TURKEY

C. Ek
University of Liege Dept. of Non-
Ferrous Metallugy and Mineral
    Dressing,
Rue Armand Stevart 2,
Bk-4000 Liege, BELGIUM

Gundiler, I.
Metallurgy Dept., New Mexico
Institute of Mining and Technology
Socorro, NM 87801, U.S.A.

Gunduz, I.
M.T.A. Teknoloji Dairesi Cevher
Zenginlestirme Servisi
Ankara, TURKEY

Hacioglu, E.
Etibank Ergani Bakir Isletmesi
Maden, Elazig, TURKEY

Hosten, C.
Mining Eng. Dept., Middle East
Technical University,
Ankara, TURKEY

Ipekoglu, B.
Cekmece Nuclear Research Center
PK 1 Havaalani, Istanbul, TURKEY

Karahan, S.
Etibank, Guneydogu Anadolu
Fosfatlari Grup Baskanligi
Ankara, TURKEY

Kelland, R.*
Applied Magnetism Group
Francis Bitter National Magnet
    Lab.
Massachusetts Inst. of Technology,
Building NW 14
Cambridge, MA 02139, U.S.A.

Matis, K.A.
Laboratory of General and
Inorganic Chemical Technology
(114), Aristotelian University
of Thessaloniki, GREECE

Oran, B.
Kure Bakirli Pirit Isletmesi
Kure, Kastamonu, TURKEY

Ozbayoglu, G.*+
Mining Eng. Dept. Middle East
Technical University
Ankara, TURKEY

Ozbayoglu, S.
Sair Hedim Sok. 2/9
A. Ayranci
Ankara, TURKEY

Ozkanli, A.
Etibank Proje Thesis Dairesi
Ankara, TURKEY

Ozturk, N.
Etibank Bursa maden Isletmeleri
Muessese Mudurlugu, Uludag
    Flotasyon
Bas Muhendisi,
Bursa, TURKEY

Parker, M.R.*
Separation Science Group
Dept. of Pure and Applied
Physics, University of Salford
Calford M5 4 WT, ENGLAND

Petruk, W.*
Research Scientist, CANMET
555 Booth St., Ottawa, Ontario
CANADA K1A 0G1

Raghavan, S.*
Dept. of Metallurgical Eng.
University of Arizon
Tucson, AZ 85721, U.S.A.

Ragot, J.*
Centre de Recherche et de
Valorisation des Minerais
Rue de Doyan Marcel Roubalt
BP 40 54201 Vondoeuvre
Cedex, FRANCE

Rocha E Silva, J.A.C.
Departamento de Minas Faculdade
de Engenharia Do Porto Roa Dos
Bragas 4088 Porto Codex, PORTUGAL

Saglam, S.
Etibank Proje Tesis Dairesi
Ankara, TURKEY

Schurger, M.L.*
Garrett Turbin Engine Co.
111 S. 34th St.
Pheonix, AZ 85010, U.S.A.

Smith, H.W.*
Chairman, Dept. of Electrical Eng.
University of Toronto
Toronto, CANADA M5 IA4

Smith, M.R.,*
Imperical College of Science
and Technology,
Dept. of Mineral Resources Eng.
royal School of Mines
London S.W.7 2BP ENGLAND

Tanrikulu, B.
Etibank Proje Tesis Dairesi
Ankara, TURKEY

Tolun, R.*
I.T.U. Kimya Metalurji Fakultesi
Kimya Muhendisligi Bolumnu
Ayazaga Kampusu
Istanbul, TURKEY

Tuzun, A.
M.T.A. Teknoloji Dairesi Cevher
Zenginlestirme Servisi
Ankara, TURKEY

Uzun, H.
M.T.A. Teknoloji Dairesi Cevher
Zenginlestirme Servisi
Ankara, TURKEY

Wills, B.A.
Camborne School of Mines Pool
Redruth Cornwall
TR 15 35E, ENGLAND

Yarar, B.*+
Metallurgical Engineering Dept.
Colorado School of Mines
Golden, CO 80401, U.S.A.

Cumhur, Y.
Etibank Guneydogu Anadolu
   Fosfatlar
Grup Baskanligi
Ankara, TURKEY

Yucel, O.
Dept. of Civil Eng.
Virginia Polytechnic Institute
and State University
Balcksburg, VA 24061, U.S.A.

*: Lecturer
ganizing Committee

AUTHOR INDEX

356

358

SUBJECT INDEX